Building the Canal to Save CHICAGO

Richard Lanyon

First Edition

Building the Canal to Save Chicago
Richard Lanyon

Published 2011 by Richard Lanyon
Distributed by:

Lake Claremont Press: A Chicago Joint, an imprint of Everything Goes Media, LLC
www.lakeclaremont.com
www.everythinggoesmedia.com

Copyright © 2016 by Richard Lanyon

All rights reserved. No part of this book may be reproduced or transmitted in any form or by any means, electronic or mechanical, including photocopying, recording, or by any information storage or retrieval system without written permission from the publisher, except for the inclusion of brief quotations in a review.

20 19 18 17 16   10 9 8 7 6 5 4 3 2

To engineers for transforming ideas
into working systems

To operators and trades for ensuring
continued service of systems

Contents

List of Maps IX
List of Photographs X

PREFACE XVII

PROLOGUE IXX

CHAPTER 1: PROBLEM AND SOLUTION 1
Problem 1
First Reversal of the Chicago River 2
Citizens' Association 4
Commission on Drainage and Water Supply 4
Recommended Solution 7
Legislative Solution 8

CHAPTER 2: ACT OF 1889 15

CHAPTER 3: FORMATION OF THE SANITARY DISTRICT OF CHICAGO 19
Petition for a Referendum 19
Drainage Boundary Commission 19
Referendum and Election of the First Board 20
Legal Challenges 20
SDC Board Gets Organized 21

CHAPTER 4: FINDING DIRECTION 25
Getting Started 25
The Chief Engineer is at Odds with the Board 26
The Board Picks a New Chief 27
Another New Chief 28
New Plan for the Main Channel 29
New Members on the Board 31
Review of Past Planning 32

Strategy for the Future 33
Another Change in Chief Engineer 34

CHAPTER 5: ROCK SECTION 39
Introduction 39
Bidding on Alternative Channel Schemes 40
Award of the First Contracts 42
Contract Requirements 43
Shovel Day: September 3, 1892 48
Change in Grade of the Channel and the Fifth Chief Engineer 49
River Diversion Channel, Goose Lake, and the Levee Height 50
Retaining Walls 51
Conglomerate Rock in Glacial Drift 52
Last Section 52
Setting of the Tablet 53
Controlling Works at Lockport 54
Bridges 57
Pumping of Drainage 57
Earthquake Damage 58
Contractor Continuity and Completion of the Work 59
Construction Progress and Methods 60
1893–1899 Construction Seasons 62

CHAPTER 6: EARTH AND ROCK SECTION 117
Introduction 117
Route Selection and Design 118
Contract Requirements 119
Des Plaines River Spillway 121
Summit-Lyons Conduit and Levee 121
Diversion Channel and Levee Height 122
Changes in Sections A and B 122
Difficult Materials in the Glacial Drift Controversy 124
Bridges 125
Contractor Continuity and Completion of the Work 126

Construction Progress and Methods 126
1893–1899 Construction Seasons 127

CHAPTER 7: EARTH SECTION 149
Decisions on Route Selection 149
Channel Design and Contract Award 152
Collateral Channel 154
Contract Requirements.. 155
Excavating Machines 157
Bridges 159
Construction Methods and Contractor Continuity 160
1894–1899 Construction Seasons 161

CHAPTER 8: JOLIET PROJECT 185
Route Selection and Decisions.. 185
Award of Contracts 189
Canal Commissioners' Lawsuit 190
Water Power.. 191
Additional Agreements and Authorizations 192
Contract Requirements.. 193
Bridges 197
1898–1899 Construction Seasons 197

CHAPTER 9: CHICAGO RIVER IMPROVEMENT 221
Temporary Relief 222
Plans for Improvement.. 225
Permits from the Secretary of War.. 227
Contract Requirements.. 228
Bridges 230
1898–1899 Construction Seasons 231

CHAPTER 10: BRIDGES 245
Early Considerations... 245
Indecision Over the Type of Bridges 246
Swing Bridges... 249

Fixed Bridges 250
The Eight-Track Bridge 250
Santa Fe System 252
South Branch Bridges 253
Joliet Project Bridges 253
Other Bridges 253

CHAPTER 11: ANCILLARY ISSUES 311
Worker Health 311
Labor Relations 312
The Employed and the Unemployed 313
Table 1 Average Number of Men Working Each Month 314
Public Order and Police Power 314
Floods and Water Supply 316
Construction Inspection and Testing 316
Construction Verification 316
Water Quality Impact on the Illinois and Mississippi Rivers 317
Raising Revenue 317
Relations with the Federal Government 318
Relations with the I&M Canal Commissioners 319
Relations with the City of Chicago 320
Communications and Travel along the Main Channel 320
Removal of Obstructions in the Des Plaines and Illinois Rivers 321

CHAPTER 12: PLACING THE MAIN CHANNEL IN OPERATION 329
Requirements of the Act 329
Initiating the Approval Process 330
Report of the Special Commission 331
Response by the SDC 333
Opening the Main Channel 334
Closing the Books on the Special Commission 337

EPILOGUE 349

ACKNOWLEDGMENTS 353

BRIDGE LIST APPENDIX 355

GLOSSARY 361

INDEX 363

Note: A set of photographs for each chapter follows each chapter, except for chapters 2 and 3.

List of Maps

Map 1	Canals and Rivers Prior to 1892	10
Map 2	Sanitary District of Chicago Boundary, 1889 to 1903	23
Map 3	Stages of Work to Reverse the Flow of the Chicago River	35
Map 4	Alternate Routes of the Main Channel from Chicago to Summit in 1891	36
Map 5	Rock Section of the Main Channel	68
Map 6	Change in Limit of Sections 1 and A of the Main Channel	69
Map 7	Section 8 of the Main Channel	70
Map 8	Main Channel Terminus and Lockport Controlling Works	71
Map 9	Earth and Rock Section of the Main Channel	131
Map 10	Sections F and G of the Main Channel	132
Map 11	Earth Section of the Main Channel	165
Map 12	Sections N and O of the Main Channel	166
Map 13	Joliet Project	200
Map 14	Section 16 of the Joliet Project	201
Map 15	Sections 17 and 18 of the Joliet Project	202
Map 16	Chicago River Improvement	233

References

Atlas of the World. Railway Terminal Map of Chicago. Chicago: Rand, McNally & Co., 1896.

Map of the Illinois and Des Plaines Rivers from Chicago to Alton, US Army Corps of Engineers, 1902–1905, Sheet Nos. 12 and 13.

Randolph, Isham. *The SDC and the Chicago Drainage Canal*. SDC, 1909.

SDC. ROW of the Main Channel. Engineering Department.

List of Photographs

CHAPTER 1: PROBLEM AND SOLUTION
1.1 Home on the I&M Canal 11
1.2 I&M Canal at Summit-Lyons Road 12
1.3 I&M Canal Near Lemont 13
1.4 Chicago Terminal Railroad Bridge over South Branch 14

CHAPTER 4: FINDING DIRECTION
4.1 SDC Board of Trustees Meeting 37
4.2 Stephens Street in Lemont 38

CHAPTER 5: ROCK SECTION
5.1   Levee between Main Channel and Des Plaines River in Section 6 72
5.2   Hydraulic Dredge Used in Section 6 73
5.3   Manual Labor Loading Overburden in Section 2 74
5.4   Steam Shovel in Section 5 75
5.5   Steam Shovel in Section 15 76
5.6   Exposed Bedrock Surface in Section 5 77
5.7   Channeling Machine 78
5.8   Rock Excavation in Section 11 79
5.9   Loading Broken Rock in Hopper 80
5.10 Double Leg Derrick in Section 14 81
5.11 Cantilever Incline Truss Conveyor in Section 12 82
5.12 Removal of Rock in Lifts under Cantilever Inclines in Section 12 83
5.13 Removal of Rock in Lifts under Cableway in Section 8 84
5.14 Incline up Vertical Rock Wall in Section 9 85
5.15 Dynamite Blast in Section 10 86
5.16 Nearly Completed Main Channel and I&M Canal in Section 11 87
5.17 Failed Wall in Section 13 88
5.18 Cableway Tower Collapse in Section 8 89
5.19 Masonry Wall Failure in Section 5 90
5.20 Rock Wall Failure in Section 5 91
5.21 Wall Separating Contract Sections 7 and 8 92

5.22	Santa Fe Railroad Crossing in Section 8	93
5.23	Santa Fe Railroad Crossing during Bridge Construction in Section 8	94
5.24	Removal of Rock under Santa Fe Railroad Bridge in Section 8	95
5.25	Masonry Wall Construction in Section 5	96
5.26	Masonry Wall Construction in Section 4	97
5.27	Masonry Wall Construction in Section 4	98
5.28	Concrete Wall Construction in Section 14	99
5.29	Tablet Day: September 3, 1895, in Section 10	100
5.30	Granite Tablet in 1899	101
5.31	Spectators at the Main Channel Transition near Willow Springs	102
5.32	Main Channel Transition from Section 1 to Section A	103
5.33	Pumping Plant for Dewatering the Main Channel Excavation	104
5.34	Section 15 Turning Basin and Lockport Controlling Works Sluice Gates	105
5.35	Lockport Controlling Works Fabricated Sluice Gates	106
5.36	Lockport Controlling Works Sluice Gate Pier Construction	107
5.37	Bear Trap Dam Foundation Excavation	108
5.38	Bear Trap Dam Gate Recess Completed	109
5.39	Bear Trap Dam Upstream and Downstream Gate Leaves	110
5.40	Bear Trap Dam Downstream Leaf Detail	111
5.41	Bear Trap Dam Leaves Completed with Hoists	112
5.42	Bear Trap Dam Inspection Crowd	113
5.43	Bear Trap Dam Completed	114
5.44	Lockport Controlling Works Panorama, Left Side	115
5.45	Lockport Controlling Works Panorama, Right Side	116

CHAPTER 6: EARTH AND ROCK SECTION

6.1	Hydraulic Dredge in Section A	133
6.2	Dipper Dredge in Section A	134
6.3	Mule Drawn Scraper in Section D	135
6.4	Excavation in Section F	136
6.5	Steam Shovel in Section A	137
6.6	Incline Track to Spoil Area in Section A	138
6.7	Steam Power Plant and Truss over Spoil Area in Section A	139

6.8	Steam Shovel in Section D	140
6.9	Boulders in Section B	141
6.10	Pumping Plant in Section B	142
6.11	Slope Paving in Section F	143
6.12	Section E Excavation near Completion	144
6.13	Transition in Main Channel Capacity from Sections F to G	145
6.14	Construction of the Des Plaines River Spillway	146
6.15	Completed Des Plaines River Spillway	147

CHAPTER 7: EARTH SECTION

7.1	Steam Shovel in Section G	167
7.2	Conveyor and Incline in Section G	168
7.3	Trusses over Spoil Pile in Section G	169
7.4	Mason-Hoover Excavating Machine in Section H	170
7.5	Damaged Mason-Hoover Machine in Section H	171
7.6	Excavation in Section I	172
7.7	Steam Plant and Incline in Section I	173
7.8	Excavation in Section L	174
7.9	Excavation Nearly Complete in Section L	175
7.10	Spoil Piles along the Earth Section	176
7.11	Steam Shovel in Section M	177
7.12	Incline in Section M	178
7.13	Spoil Pile in Section M	179
7.14	Dipper Dredge in Section O	180
7.15	Dipper Dredge in Section O	181
7.16	Dipper Dredges in Section O	182
7.17	Tugboat Used in Section O	183
7.18	Excavation Completed in Section O	184

CHAPTER 8: JOLIET PROJECT

8.1	Des Plaines River Excavation in Section 16	203
8.2	Des Plaines River Excavation in Section 16	204
8.3	Des Plaines River Excavation in Section 17	205
8.4	Des Plaines River Excavation in Section 17	206

8.5	Des Plaines River and I&M Canal in Joliet	207
8.6	Des Plaines River Excavation in Section 18	208
8.7	Des Plaines River Excavation in Section 18	209
8.8	I&M Canal in Section 18	210
8.9	Reconstruction Work on Lock No. 5	211
8.10	Reconstruction Work on Lock No. 5	212
8.11	Reconstruction Work on Lock No. 5	213
8.12	Dam No. 1 before Reconstruction	214
8.13	Rebuilt Lock No. 5	215
8.14	Navigation Tow in Lock No. 5	216
8.15	Work on New Powerhouse	217
8.16	New Dam No. 1	218
8.17	Work on New Powerhouse	219

CHAPTER 9: CHICAGO RIVER IMPROVEMENT

9.1	Preconstruction Building Survey	234
9.2	Work on the Upstream End of the Bypass Channel	235
9.3	Work on the Bypass Channel South of Adams Street	236
9.4	Work on the Bypass Channel at the West Side Elevated Railroad Bridge	237
9.5	Work on the Bypass Channel South of the West Side Elevated Railroad Bridge	238
9.6	Steel Framing for the Bypass Channel	239
9.7	River Wall Cofferdam for the Bypass Channel	240
9.8	Dipper Dredge in the South Branch Used in the Construction of the Bypass Channel	241
9.9	Interior of the Bypass Channel from the South End	242
9.10	Interior of the Bypass Channel from the North End	243
9.11	Entrance to the Bypass Channel Upstream of Jackson Street	244

CHAPTER 10: BRIDGE CONSTRUCTION DETAILS

10.1	Quarried Rock Used for Abutments, Foundations, and Piers	255
10.2	Steel Delivered for Bridge Superstructure Erection	256
10.3	Temporary Timber Trestle Bridge Construction	257
10.4	Temporary Timber Trestle Bridge Construction	258

10.5	Wooden Pile Foundations	259
10.6	Wooden Pile Foundations	260
10.7	Abutment and Center Pier Construction	261
10.8	Abutment Construction	262
10.9	Abutment Foundation Excavation	263
10.10	Abutment Foundation Excavation	264
10.11	Abutment Construction	265
10.12	Swing Pier Foundation Preparation	266
10.13	Swing Pier Turntable Installation	267
10.14	Bobtail Swing Bridge Erection	268
10.15	Bobtail Swing Bridge Test	269
10.16	Center Pier Swing Bridge Superstructure Erection	270
10.17	Center Pier Swing Bridge Superstructure Erection	271
10.18	Center Pier Swing Bridge Superstructure Erection	272
10.19	Center Pier Swing Bridge Superstructure Erection	273
10.20	Center Pier Swing Bridge Superstructure Erection	274
10.21	Center Pier Swing Bridge Superstructure Erection	275
10.22	Center Pier Swing Bridge Superstructure Erection	276
10.23	Superstructure Erection Completed	277
10.24	Fixed Girder Placement	278
10.25	Jacking Up Truss to Raise Piers	279
10.26	Truss Collapse Due to Shifting False Work	280
10.27	Truss and Pier Raising Compete with Added Truss	281
10.28	Raising Railroad Track Grade	282
10.29	Street Viaduct Construction	283

CHAPTER 10: COMPLETED BRIDGES OVER THE MAIN CHANNEL

10.30	Southwestern Boulevard/Western Avenue Bridge in Section O	284
10.31	Interim Eight-Track Railroad Bridges in Section O	285
10.32	Chicago, Madison, and Northern Railroad Bridge in Section O	286
10.33	Kedzie Avenue Bridge in Section N	287
10.34	Santa Fe Railroad Bridge in Section N	288
10.35	Belt Line Railroad Bridge in Section K	289
10.36	Santa Fe Railroad Bridge in Section G	290

10.37 Summit-Lyons Road Bridge in Section F 291

10.38 Chicago Terminal Railroad Bridge in Section E 292

10.39 Willow Springs RoadBridge in Section 1 293

10.40 Santa Fe Railroad Bridge in Section 8 294

10.41 Stephens Street/Lemont Road Bridge in Section 8 295

10.42 Western Stone Company Railroad Trestle in Section 9 and 10 296

10.43 Romeoville Road Bridge in Section 12 297

CHAPTER 10: COMPLETED BRIDGES OVER THE DES PLAINES RIVER

10.44 Santa Fe Railroad Adjacent to Section F 298

10.45 Summit-Lyons Road Adjacent to Section F 299

10.46 Chicago Terminal Railroad Adjacent to Section E 300

10.47 Santa Fe Railroad Adjacent to Section 8 301

10.48 Lemont Road Adjacent to Section 8 302

10.49 Lockport (Ninth Street) Road in Section 16 303

10.50 Wire Mills (Sixteenth Street) Road in Section 16 304

10.51 Elgin, Joliet, and Eastern Railroad in Sections 16 and 17 305

10.52 Cass Street in Section 18 306

10.53 Jefferson Street in Section 18 307

10.54 Chicago, Rock Island, and Pacific Railroad in Section 18 308

10.55 Stephen Street Viaduct under the Santa Fe Railroad
       Adjacent to Section 8 309

CHAPTER 11

11.1 Police Officers and Staff outside of a Field Office 323

11.2 Engineering Staff outside of a Field Office 324

11.3 Compressed Air Plant and Construction Camp 325

11.4 Construction Camp 326

11.5 House or Shelter Made of Rock Spoil 327

11.6 Visitors in Hopper Suspended from a Cableway 328

CHAPTER 12: PLACING THE MAIN CHANNEL IN OPERATION

12.1 Location Where Water Will Enter for Filling
      the Main Channel Excavation 339

12.2 Main Channel Excavation Is Ready for Filling 340

12.3	Initial Inflow of Water for Filling the Main Channel Excavation	341
12.4	Witnesses to the Filling of the Main Channel Excavation	342
12.5	Removing the Earthen Plug to Allow Water to Fill the Main Channel	343
12.6	Rising Water Level at the Lockport Controlling Works	344
12.7	The Main Channel Is Full of Water	345
12.8	The Water at the Bear Trap Dam Is Awaiting the Approval to Discharge	346
12.9	Discharge of Water at the Bear Trap Dam Reversing the Flow in the Chicago River	347

All photographs courtesy of the Metropolitan Water Reclamation District of Greater Chicago. Cover from MWRD photograph, disk 14, image 26. Photographs 6.14 and 6.15 undoubtedly originated with the MWRD but were not found in the MWRD archive.

Please note that due to age and storage conditions over the years, some of the glass plate negatives were damaged or faded. The collection now resides under proper storage conditions in the Illinois State Library Archives in Springfield. Also, negatives for the Geiger set do not exist with the MWRD. Prints available in the MWRD archive were used. Permission to use these photographic images and the assistance of the MWRD are acknowledged and appreciated.

Preface

Nearly all my life I've had contact with the water in and around Chicago. I grew up and attended schools along the North Branch, often played at Montrose beach on Lake Michigan, began college at the Navy Pier campus of the University of Illinois and worked as a research assistant on a Lockport Powerhouse model study while at the Urbana campus. After receiving my masters of science degree in civil engineering and water resources, I joined the Harza Engineering Company in Chicago and worked on Great Lakes diversion studies. In 1963 I began my nearly forty-eight-year career at the Metropolitan Sanitary (now Water Reclamation) District of Greater Chicago (MWRD), beginning as an associate engineer in the Waterways Control Section.

Preparing this history and description of a landmark public works project has been both a labor of love and a rewarding experience. It began in the 1990s as the Chicago Sanitary and Ship Canal approached a century of service and it was my intent to have this history ready for the centennial on January 17, 2000. I spent hours reviewing historical documents and virtually read the entire 7,000-page Proceedings of the Board of Trustees of the Sanitary District of Chicago for the 1890 through 1900 period.

In 1998, I was struggling with the MWRD archive of photographs on glass plate negatives. Only recently had work begun to electronically scan the negatives and make them available for viewing in a more user-friendly manner. The first ten or more years of negative scans were available and perhaps a thousand images were viewed. The problem was to make a limited selection to amplify the written history with visual impressions. In 1999 my life changed as I assumed more demanding responsibility and the need to pursue important initiatives with my promotion to director of research and development (now research and monitoring) for the MWRD. It was necessary to lay the work on this history aside and forego the goal of beating the centennial deadline.

The first decade of the twenty-first century provided no relief from a demanding schedule of MWRD work obligations and with my appointment as general superintendent (now executive director) of the MWRD in June 2006, I thought that I might never be able to finish this history. With retirement at the close of 2010, it became my new goal to finish this work in 2011.

Nearly fifty years of professional experience with water in the metropolitan Chicago area has given me a fundamental and detailed understanding of the entire public plumbing system, both local and national. All aspects of water in our lives are included: water source, water supply, sewerage, drainage, reclamation, stormwater, flooding, watersheds and the rivers that return to the oceans. Without this infrastructure, which we all take for granted, Chicago may not have survived the limitations of the natural systems to become the vibrant metropolis it is. The man-made canal system, the backbone of which is the Chicago Sanitary and Ship Canal, will continue to sustain metropolitan Chicago.

The history of this little understood canal system is important as we contemplate changes. With the flat topography of the Chicago region, any disturbance of the surface drainage system may have profound effects on our future welfare.

The technically oriented reader will enjoy reading the details of the engineering and construction challenges in building the canal and many bridges. Those who may feel technically challenged will enjoy the photographs and captions. The rich archive of photographs is a story in itself.

Prologue

"So they reversed the river." With these words, one of the world's most remarkable engineering public works projects is often summarized. Representatives of the Metropolitan Water Reclamation District of Greater Chicago frequently speak to civic and school groups about the construction of the canal that saved Chicago from a perilous nuisance and public health problem. Solving this problem enabled Chicago to grow into a world-class city. As time passes, details become obscure. So after more than a century, it is understandable how it is possible to compress this remarkable engineering feat into the five simple above quoted words. This book *Building the Canal to Save Chicago* provides the detail.

On January 17, 2012, the Main Channel observed 112 years of service, well over a century of continuous operation. Its conception, design, and construction, as well as the organization responsible, is a story worth telling. It has not been told in the depth necessary to understand the technical details; the planning and design decisions; the complications in construction; the difficulties, delays, and failures of contractors; the resolution of conflicting interests; and the knowledge that, after all is done, it would work.

Chicago raised itself out of the mud to overcome the poor drainage of its flat topography. It survived a devastating fire and rebuilt itself. Now, the dilemma of a determined city wanting to grow but faced with its water supply coming from its toilet, a technical solution was the first need. Next, it was necessary to find the political mechanism for implementation. Last was the long, grinding, methodical progress of putting a public works project in place. Will it work? How to do it? Do we have the technology? How to raise the revenue?

The story proceeds from the problem the city faced through to the opening of the channel. It includes finding the technical and political solution, organizing a public agency and putting it into operation, preliminary planning and design decisions, contracting and construction problems and solutions, and at last, putting the channel into operation. Not just a civil works project, this undertaking also involved social issues regarding labor, health, and sanitation of the workforce and maintaining public order.

This project was conceived and construction was nearly completed as federal authority over waterways was emerging. Just before the project was completed and operation began, local leaders were faced with new challenges from the federal government that could have put the project on hold. More importantly, if the challenge wasn't firmly and strongly confronted, the growth of the city would have been impaired. It probably wasn't the first, and it certainly wasn't the last, example of local advocacy in implementing a public works project, pitting the determination and defiance of a local authority with a critical state mandate against a federal department finding its mission.

Litigation to stop the canal was initiated by a downstream city who felt threatened by an upstream city's wastewater. This quickened the action by local leaders to finish the project. The litigation was stalled by politics and other priorities and never derailed Main Channel completion.

The Proceedings of the Board of Trustees of the Sanitary District of Chicago (SDC) is the primary source of this story. Other sources are found at the conclusion of each chapter. This is simply the story of the complexity of implementing a massive, and innovative for the time, public works engineering project.

More color and controversy could be included if one were to follow accounts in the news media, but this is not the intent of this telling. If interested, the reader is encouraged to read other accounts found in *The Chicago River: A Natural and Unnatural History* by Libby Hill or

The Lost Panoramas: *When Chicago Changed Its River and the Land Beyond* by Michael Williams and Richard Cahan. These authors have developed their history from the popular media and other noteworthy sources.

Any public works project will invite opposition and criticism, and the undertaking can be made to look foolish and ill conceived at times. The story at hand is no different. All that has passed and what remains is the evidence of success: a city that went on to grow and prosper because of the canal system conceived, financed, and implemented by local leadership. After more than a century of service the canal system continues to provide economy and efficiency in water transportation, protect public health and welfare by efficient drainage of treated wastewater and urban storm water, offer an amenity for waterfront property, allow reuse of water for industrial cooling, and be an outlet for those people seeking water recreation in an intense urban setting.

Chapter 1

Problem and Solution

Chicago developed in a location not well suited for a high concentration of people and industry. The topography was flat, poorly drained, and wet during most of the year. People and industry were interdependent, but the natural setting was not conducive to either or both in large numbers.

Problem

From the middle to the end of the nineteenth century, the rapidly growing city of Chicago was plagued with frequent epidemics of waterborne diseases. The city's death rate was one of the highest of major cities in the world. Despite the introduction and widespread use of sewers beginning in 1855, the problem did not abate. Typhoid alone claimed fifty lives per one hundred thousand population on average for the last three decades. In fact, the sewers, while they provided adequate local drainage, facilitated the public health nuisance since they conveyed sewage to the river with greater haste. This resulted in polluted water being discharged to Lake Michigan, the only outlet of the Chicago River, with greater speed and in larger volume.

Lake Michigan was also Chicago's water supply; thus, the city was in a vicious cycle. Successive efforts were undertaken to move the water intakes farther from shore. But in time, the plume of pollution would reach farther out into the lake, enveloping the intakes. At this time, technologies for potable-water treatment and sewage treatment were not practiced or proven.

First Reversal of the Chicago River

In 1848, the I&M Canal was opened to navigation, connecting Lake Michigan to the Illinois River through the Chicago River, over the former Chicago Portage, and down the Des Plaines River. This canal allowed a small amount of the polluted Chicago River water to flow away from the lake. However, in wet years, the capacity of the I&M Canal was insufficient to remove all pollution. In times following a storm, the floods on the Chicago River were too great to allow the I&M Canal to provide any degree of relief. Chicago increased the capacity of the I&M Canal with a pumping station at the canal entrance near Bridgeport in 1871, but this provided only marginal improvement during dry years. It was insufficient during wet years and storms.

Some thought that the I&M Canal should provide the needed relief if only the canal operators would open the lock gates at Lockport, the first lock after Bridgeport. This thought was a popular misconception regarding the hydraulic capacity of this long canal. It is not unlike the misconception in present times that opening the Chicago River lock will prevent basement flooding in neighborhoods removed from the waterway. However, the engineers knew only too well of the limited capacity of the I&M Canal. The pumping station at Bridgeport was primarily for providing adequate flow for navigation when the lake and river level were low and for diverting as much wastewater as possible. However, pumping too much water into the I&M Canal at Bridgeport would only result in overtopping the canal banks and causing the excess water to flow back to the South Branch through the West Fork (see map 1).

Another problem was the Des Plaines River, which in wet weather would go out of its banks north of Summit and discharge part of its excess toward Chicago. This overflow was a natural outlet for the river bequeathed upon the landscape by the glacial sculpting of the land surface. It was what the early explorers observed to be a potential water connection, later named the Chicago Portage, between the Great Lakes and the Mississippi River. Mud Lake, an area of open water in wet years, was along the route of the portage.

Two of Chicago's early mayors were into land and development speculation. If Mud Lake could be drained, a vast area could be opened for farming and housing. Even commercial or industrial development was possible, given the proximity to the I&M Canal and railroads. The Ogden-Wentworth Ditch was constructed about 1870 to provide drainage for Mud Lake. This ditch connected on the east to the West Fork and on the west to the Des Plaines River, providing drainage to the east since Lake Michigan was a lower outlet for the Chicago River and its network of branches and forks. The Des Plaines River downstream of Summit was of flat gradient, heavily vegetated, and sluggish in flow. On occasion the Des Plaines River would overtop the bank of the I&M Canal, causing bank erosion and sediment to be deposited in the canal. The new ditch provided another outlet for the Des Plaines River and abetted the overflow of Des Plaines River water toward Chicago.

To prevent Des Plaines River floods from damaging Chicago, the city built the Ogden Dam and embankments near the entrance to the Ogden-Wentworth Ditch. The embankments channeled the river flow from going overland, and the dam was intended to provide beneficial low flows from the Des Plaines River to cleanse and flush the West Fork and South Branch. However, being remote from the city, maintenance of the dam was neglected and it was easily vandalized. Too often, the flow toward Chicago increased and was less a benefit and more a problem. Excess flow from the Des Plaines River coursing through the Ogden-Wentworth Ditch eroded the ditch and carried excessive sediment. The sediment would be picked up by the Bridgeport Pumping Station and ended up being deposited in the I&M Canal.

There were two ways to deal with the excess flows of the Des Plaines River. Either enlarge its capacity downstream of Summit to provide for all floodwaters or divert excess floodwaters to the lake through a controlled channel north of the city. The latter was studied extensively up to the 1890s, but it was never implemented. Expert opinions were divided on the solution of the Des Plaines River excess flows. However, for positive reversal of the Chicago River when the lake level was low, there was general consensus: the only way to provide relief was through the construction of a much larger and deeper channel than the I&M Canal.

Citizens' Association

In 1880, the Citizens' Association took upon itself the task of investigating alternative remedies and broadcasting these to residents and policy makers. It was a technical issue to provide the proper fix, but social action was necessary to prod the policy makers. The affluent could afford bottled water brought in from natural springs or from expensive groundwater wells. But to the common people, whose labors made Chicago grow and become an industrial and commercial center, bottled water was unaffordable. The sewage disposal issue had to be successfully and completely solved, or Chicago would not continue to grow and prosper into the twentieth century.

It also helps to have a catastrophe to demonstrate and promote the need for action. Unfortunately for those that suffered loss from the storm, but fortunately for the greater and long-term public good, Mother Nature cooperated on August 2, 1885. A large storm over the area once again caused the Des Plaines River to go out of its banks near Summit and send a torrent of water eastward along the Ogden-Wentworth Ditch to the West Fork of the South Branch. The flood wave went up the branch and out the Chicago River main stem into Lake Michigan, causing extreme damage to boats, docks, and bridges. Bridges with narrow and restrictive waterway openings were susceptible to damage.

Fortunately, the storm brought a cool front and onshore wind. The flood wave purged the river of offensive deposits, the onshore winds kept the plume of pollution away from the water intakes, and the cool weather kept the odors in check. Were conditions otherwise, a greater catastrophe might have ensued.

Commission on Drainage and Water Supply

Both the citizenry and news media were clamoring for action. In late August, the Citizens' Association issued a report and called for action, seeking a permanent solution to the sanitary problem. Among the authors of this report was Lyman E. Cooley, who would later be instrumental in the construction of the channel to reverse the river. The city of Chicago's Common Council responded in January 1886 with the formation of the Commission on Drainage and Water Supply. This action was the first official movement by the city toward a major undertaking to solve the problem of the Chicago River. Mayor Carter H. Harrison appointed Rudolph Hering of Philadelphia as chief engineer of the commission and Benezette Williams and Samuel G. Artingstall, city engineer, as consulting engineers to the commission. The latter two would later be involved in the engineering design of the channel to reverse the flow in the river. The commission was charged with the task of outlining a solution to the problem by January 1887, in time for authorization by the Illinois legislature.

The commission issued its thirty-six-page preliminary report with figures and tables in January 1887, outlining three alternatives and recommending one. From April through December, the commission staff gathered information, reviewed the work of others, conducted topographic surveys, performed hydrographic and hydrologic studies, and prepared preliminary designs and cost estimates. The staff included Lyman E. Cooley, mentioned above, and T. T. Johnston. Both were engineers who would eventually be involved in the design and construction of the channel to reverse the flow in the river. The report discussed the current sewage disposal system and population projections. Three alternatives for solving the problem were discharge of sewage into Lake Michigan, disposal of sewage on land, and discharge of sewage into the Des Plaines River.

Discharge to the lake would involve large conveyance systems to bring fresh water from north of Evanston into the city and to convey waste south to the lake in the Calumet region. Dispersion of the waste in the lake and removal by the prevailing currents were speculated to be sufficient, yet waste treatment was considered an eventuality and judgment was reserved on the long-term impact on the lake. Disposal on land was only touched upon briefly, noting that it appeared to be working satisfactorily in the Pullman area but that the amount of land, the size of conveyance and distribution systems, and the extent of underdrainage and land preparation for a large city would be cost prohibitive. Discharge to the Des Plaines River needed little explanation due to the extent of existing studies and plans. Attention was devoted to the quantity of water needed for dilution, circulation requirements in various branches of the rivers, schemes for diverting flood flows on the Des Plaines and North Branch to reduce flood flows through the city, and the possible routes of a channel to reverse the flow in the Chicago River.

Cost estimates were based on a projected population of 2,500,000, roughly three times the then-current population. Discharge to the lake was estimated at $37,000,000, disposal on land at $58,000,000, and discharge to the Des Plaines River at $28,000,000. Short-term projects for relief of the lake and the Chicago River were recommended. These included diversion of flood flows in the North Branch and Des Plaines, improvement of the Fullerton Avenue pumping station, redirecting sewers away from the lake to the river, and enlarging the capacity of the I&M Canal to Summit with a pumping station to the Des Plaines River. Discharge to the Des Plaines River was obviously the recommended solution.

The report included extensive consideration on the water supply for the city. At the time, the city was at its limit of capacity to deliver potable water, and a new water tunnel was under construction. First, the commission studied the existing distribution system and recommended improvements to increase the efficiency of delivery. Next, they recommended additional intakes along the lake front and pumping stations throughout the city to reduce pumping cost and equalize pressure in the distribution system. Last, they recommended that the city should plan for its water system on the basis of 150 gallons per day per capita. The city has followed all these recommendations.

The report concluded with suggestions for a single management authority for the drainage systems and another for water supply. Also identified were the two drainage systems, one north of Eighty-Seventh Street with its outlet at Summit on the Des Plaines River. The other was south of Eighty-Seventh Street with its outlet at Sag, also on the Des Plaines River. Not stated, but implied by the common outlet, was one drainage authority. In his transmittal of the report to the city council, Mayor Harrison called for legislation to create a new metropolitan district to fund and prosecute the work. He suggested that it be under the direction of one able paid head, perhaps appointed by the governor, rather than a body of commissioners. Most of these suggestions have been implemented.

Recommended Solution

Regarding the recommended plan for sewage disposal, the commission set the parameters for the design of a large channel from Chicago to Joliet, using as input much of the previous plans as well as their own creative ideas and the results of their exhaustive investigations. Three criteria for adequate capacity were fundamental: storm flow, sanitation, and navigation. The storm flow capacity recognized rapid runoff from impervious surfaces and the need to prevent backflows to Lake Michigan. Sanitation capacity recognized the purification afforded by the oxygen in a large quantity of diluted water. Navigation capacity recognized the need to pass large vessels with manageable velocities. The commission recommended a channel size of 3,600 square feet in area and a velocity of 3 feet per second, which was rounded to 600,000 cubic feet per minute (cfm) to serve a population of 2,500,000 people.

It is not known what dilution factor the commission used to come up with the 600,000 cfm flow rate. The preliminary report indicated that the dilution factor would be documented in the final report, based on the work of the state board of health. In retrospect, it calculates out to about 2,600 gallons per person each day. The per capita daily water consumption in 1890 was 74 gallons, which comes out to a gross dilution factor of almost 35 gallons of dilution water to 1 gallon of water consumed. In calculating this factor, the commission must have given some consideration for the water used by industry as well as people. The commission's preliminary report was its only report. The final report intended for July 1887 was never issued.

The commission estimated the cost as ranging between $20,300,000 and $24,500,000 and recognized the potential for recovering the energy in the water flowing down the steep descent of the Des Plaines River through waterpower development. It was the commission's opinion that the large channel capacity would be sufficient to prevent storm water from flowing back into the lake. Further, it was their opinion that water-intake cribs located 2 miles offshore would safely yield abundant quantities of pure drinking water uncontaminated by storm water. Finally, the commission recommended that in order to achieve the best results, the responsibility for water supply and drainage should be placed under one management entity. Fifty years later it would be determined that a control structure at the mouth of the river would be needed to prevent storm water flows to the lake. Time would prove that safe water was available two miles from shore. Regretfully, the last recommendation was not adopted, and today the region is beset with multiple water management authorities.

Legislative Solution

The scope of the recommendations and the size of the undertaking required action by the state legislature to authorize the formation of a new unit of local government. Why was it necessary to form yet another governmental agency to accomplish the task at hand? Money and control! Chicago was in debt to the legal limit, and the task at hand would require considerable resources. A new entity encompassing a larger area could borrow anew, could have a larger tax base than the city, and would have powers beyond the city limit. The problem affected the region, not just the city, and the city would not be expected to serve more than its own interest.

Although legislation was introduced in 1885 and 1887, both attempts failed to garner sufficient votes to pass because downstate interests could not support the discharge of Chicago sewage. After the second defeat, and recognizing that something must be done, the leadership in Springfield appointed a legislative commission in May 1887, consisting of the mayor of Chicago, two members of the House, and two members of the Senate. This commission was popularly known as the Committee of Five, and its charge was to examine the problem and report at the next session. If it concluded that a waterway to receive Chicago's drainage was the most practicable solution, it was further charged with the task of defining the requirements of the waterway to protect the health and comfort of the people along the Des Plaines River at Joliet and downstream thereof. Chicago was to fund the work of the commission.

Appointed to the Committee of Five were Mayor John Roche, B. A. Eckhart, and Thomas MacMillan of Chicago; Andrew Bell of Peoria; and Thomas Riley of Joliet. Eckhart would later serve on the board of the SDC. The committee held numerous meetings in Chicago and downstate to obtain public opinion and determine what was needed for a successful outcome. It also sought technical and legal advice. Two years later, the committee reported the bill to the legislature on February 1, 1889. Although the legislation was applicable to any county in the state, it was obvious that it was designed for Chicago and for the construction of a channel to carry away Chicago's diluted sewage.

The channel was to be of ample size and capacity to dilute the sewage with sufficient quantities of Lake Michigan water, rendering the mixture inoffensive to residents along the waterway in Joliet and downstream. Few people downstate believed the added flow would be inoffensive. The channel would also be navigable to serve downstate interests. It would fulfill the dream of many for a waterway for commerce from the Great Lakes to the Gulf of Mexico right through the State of Illinois. The navigable guarantee was persuasive to downstate businesses and towns along and near the Illinois River.

References

Brown, G. P. *Drainage Channel and Waterway*. Chicago: R. R. Donnelley & Sons Company, 1894.

Chicago Daily News, August 3, 17, and 28, 1885.

Currey, J. Seymour. *Chicago: Its History and Its Builders*. Vol. 3. Chicago: S. J. Clarke Publishing Co., 1912.

Journal of the House of Representatives. Illinois: 35th General Assembly, May 26, 1887.

Preliminary Report of the Commission on Drainage and Water Supply, January 1887.

Report of the Citizens' Association, August 27, 1885.

Roche Bill reported to General Assembly, February 1, 1889; approved, May 29, 1889; effective July 1, 1889. Illinois, 1889.

SDC. Engineering Works. August, 1928.

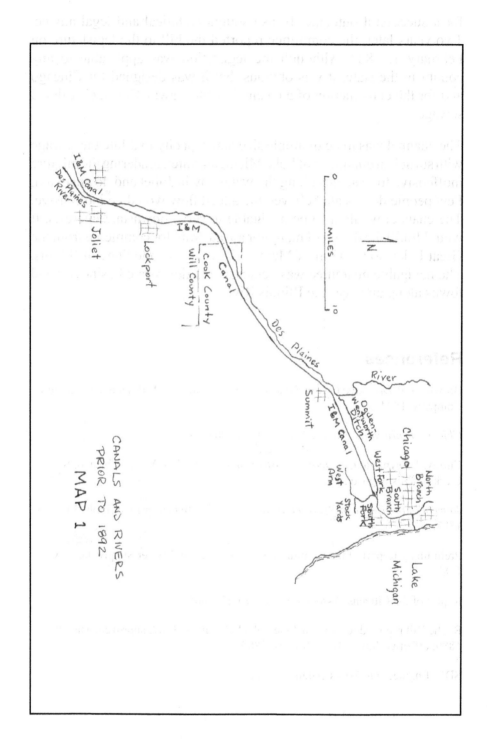

CHAPTER 1: PROBLEM AND SOLUTION

Photograph 1.1 taken on August 22, 1899, of a home on the bank of the Illinois and Michigan Canal as construction of the Main Channel was nearing completion. The I&M Canal upstream of Joliet would continue in service until about 1920 when the Calumet-Sag Channel was nearing completion. In 1907 the new lock at Lockport on the Chicago Sanitary and Ship Canal provided an alternate route for navigation between Joliet and Chicago. MWRD photo, disc 9, image 53.

11

Photograph 1.2 taken on May 28, 1894 from a point west of the I&M Canal on an embankment separating the I&M Canal from the construction of the Main Channel in the vicinity of Summit. The Summit-Lyons Road Bridge over I&M Canal is left of center. The Des Plaines River and construction of the Main Channel is to the right of the embankment. The structure in the center is believed to be a SDC canal construction field office and police station on the Summit-Lyons Road. MWRD photo, disc 3, image 10.

CHAPTER 1: PROBLEM AND SOLUTION

Photograph 1.3 taken in 1895 shows the I&M Canal near Lemont. Rock quarried from the excavation of the Main Channel to the left in this view is loaded onto flat boats for transport upstream to locations where channel walls or bridge foundations are being constructed. MWRD photo, Geiger set, image 99.

Photograph 1.4 taken on August 17, 1899 showing the Chicago Terminal Railroad Bridge over the South Branch. Swing bridges were popular, but they required a pier in the center of the river, restricting the passage of commercial vessels and causing increased flow velocity. This bridge was eventually replaced by the SDC to improve the flow conditions in the South Branch and remove obstructions to river traffic. MWRD photo, disc 9, image 46.

Chapter 2

Act of 1889

The sanitary district enabling act, officially titled "An Act to create sanitary districts and to remove obstructions in the Des Plaines and Illinois rivers," was approved on May 29 to be effective July 1, 1889. The sections of the act that are pertinent to the construction of the Main Channel are explained.

Section 1 provided for the creation of a sanitary district and its service area by referendum. A sanitary district must be contiguous and consist of at least two villages or cities, contain no territory beyond 3 miles of the village or city limits, and be limited to one county. In Section 7, authority was given for a district to construct and control channels and drains, docks on navigable channels, and water power. Districts were granted authority to acquire land by condemnation in Section 8. Under Section 9, districts may borrow money and issue bonds, but indebtedness was limited to 5 percent of the value of taxable property in the district. Interest on debt was to be paid from an annual tax, as provided in Section 10.

Section 11 required that all contracts for work by the district must be awarded to the lowest responsible bidder, persons employed by the district or its contractors must be citizens or intend to become citizens, and eight hours shall constitute a day's work. Funding for corporate purposes, provided for in Section 12, could be by a tax, the total amount of which should not exceed half a percent of the value of taxable property in the district. Section 13 provided that a special assessment could also be used to fund improvements.

Section 17 provided for the taking of public property for corporate use in the same manner as private property. Regarding the I&M Canal, a district could use the I&M Canal right of way (ROW) for its purposes at no cost, but only within the county of its jurisdiction. Such use of I&M Canal ROW remained subject to the control of the I&M Canal Commissioners.

Any channel constructed by a district for the conveyance of sewage that discharges to a river outside the district was required by Section 20 to have a capacity of at least 200 cfm for each 1,000 population tributary to the channel. The discharge of garbage and dead animals to channels was prohibited. In Section 21, the attorney general was authorized to bring action against a district for failure to comply with the statute.

Section 23 of the act specifically required any sanitary district causing Lake Michigan waters to discharge to the Des Plaines or Illinois Rivers to provide for a flow of at least 300,000 cfm in a channel having a depth not less than 14 feet and a current no greater than 3 miles per hour. Further, if the channel passes through a rocky stratum that is less than 18 feet below the level of Lake Michigan, then the channel shall have a capacity of at least 600,000 cfm, a depth not less than 18 feet, and a bottom width not less than 160 feet.

Further in Section 23, it was required that if the population that drains to the channel exceeds 1,500,000, the channel shall discharge not less than 200 cfm per each 1,000 population. When and if the state improves the Des Plaines or Illinois River to a capacity of 600,000 cfm, then the district shall, within one year, enlarge the channel leading to the Des Plaines or Illinois River to 600,000 cfm capacity. If the channel is constructed in the Des Plaines River, its capacity shall be continuous down the slope between Lockport and Joliet to the Upper Basin. The district constructing a channel to convey water from Lake Michigan may modify or remove obstructions in the Des Plaines or Illinois Rivers to prevent overflow or damage. Existing water power rights shall not be injured or destroyed.

Section 24 declared that constructed channels conveying 300,000 cfm or more are navigable and that if the channel connects to the Des

Plaines or Illinois Rivers are made navigable by the state government, the state interest shall prevail with respect to navigation, and the district shall prevail with respect to drainage and sanitation.

In Section 26, the legislation required that any municipal water supply system within a sanitary district that is protected by the sanitary district must supply water to other territories within the sanitary district that have no system and that ask for water to be supplied. Further, it must be supplied at the same cost as the cost within the municipality.

To assure that a channel constructed pursuant to Section 23 was in compliance with the section requirements, Section 27 required that the district receive approval from the governor as to the adequacy of the channel before it receives the discharge of sewage or water. The governor was to be advised by a commission of three persons appointed from downstate areas (Joliet, LaSalle, and Peoria) who would retain competent engineers to make a determination of the adequacy of the constructed channel and of compliance with the requirements of the act. Further, clarification was provided that Section 23 applied to any channel serving a population of 300,000 or more that discharges beyond the limits of the district and that any channel that receives its supply from a river connected to Lake Michigan shall be considered as receiving its supply from the lake.

In retrospect, the act was a masterpiece of legislation. It was creative and comprehensive, not only in guiding Chicago out of its half-century-long battle with the ravages of waterborne diseases and uncontrollable drainage, but also in protecting downstate residents from the potential ill effects of Chicago's sewage and in providing for a deep-water navigation link between the Great Lakes and the Gulf of Mexico.

Subsequent to the creation of the SDC under the act, the SDC saw the need to return to Springfield on numerous occasions over the years to seek amendments to the act to enable it to better serve its constituents and carry out the requirements of the law. The first such occasion was in 1893. This and other amendments are described in chapter 11.

Reference

Illinois Laws. Illinois, 1889.

Chapter 3

Formation of the Sanitary District of Chicago

Implementing the provisions of the act was no small chore. It required leadership through several steps in the organizing process.

Petition for a Referendum

The act was approved on May 29 to be effective July 1, 1889. Before these dates, however, backers of the plan were busy preparing for the rigorous task of getting the SDC underway. First was gathering over five thousand signatures on petitions to be presented to a panel of three Cook County judges who were to determine the name and boundaries of the proposed SDC. The panel was headed by Chief Judge Richard Prendergast, who would later be a member of the first Board of Trustees of the SDC and the second president of the board.

The act required the referendum to be held on the first Tuesday in November, so time could not be wasted lest another year would go by. The petitions were submitted to the panel on August 15.

Drainage Boundary Commission

The panel, known for this purpose as the Drainage Boundary

Commission, rendered its decision on October 12. The election could then proceed on November 5, as required by the law. The boundaries of the SDC included all of Chicago north of Eighty-Seventh Street, the incorporated towns of Cicero and Lyons, and a part of Lyons Township. The reason for the inclusion of Chicago is obvious. The exclusion of the area south of Eighty-Seventh Street was due to its being part of the Calumet River drainage and also due to local opposition since the area was distant from Chicago. Cicero was included because it favored inclusion. Lyons and its surrounding unincorporated area were included because of a proposed railroad yard and stockyard development. North and northwest areas were primarily devoted to farming and were therefore not included. Evanston was omitted because it was determined that the discharge of their sewage to the lake did not threaten their own, or Chicago's, water supply (see map 2).

The territory delineated was bounded roughly by the lakefront on the east, Eighty-Seventh Street on the south, Harlem Avenue on the west, and Devon Avenue on the north, a total of 185 square miles.

Referendum and Election of the First Board

The vote resulted in a landslide, epic even by Chicago proportions: 70,958 for and 242 against. Following certification of the results of the referendum by the county judge, the SDC was declared organized under the act. The county judge also called for an election of trustees for the SDC. Candidates for trustees of the SDC were placed on the ballot, and nine were elected on December 12. After certifying the election and taking the oath of office, the first meeting of the board occurred on January 18, 1890. The SDC was a reality, but there would be some legal challenges to overcome and the necessary engineering work to do before the channel to carry away Chicago's sewage would become a reality and put to use.

Legal Challenges

The challenges were two lawsuits. One, filed by Cook County, claimed that the elected trustees had no authority to hold their offices and that the sanitary district act violated the state constitution. The other was filed by a taxpayer seeking an injunction on the sale of bonds approved at the February 8, 1890, meeting to raise revenue for the expenses of the SDC. Hearings and judgments were expedited and appealed to the Illinois Supreme Court. On June 12, 1890, the court rendered its decision, finding in favor of the SDC in both cases.

In his opinion, Supreme Court Judge Bailey summarized the responsibility of the SDC as follows:

> Sanitary districts ... are unlike any class of municipal corporations heretofore existing in the State. They are ... a new product of the creative power of the Legislature. Their character and purpose are both indicated and limited by the word "sanitary," ... a sanitary district is ... organized to secure, preserve and promote the public health.

Consistent with the contemporary understanding of the methods necessary for securing and preserving the public health, the SDC's mission was to effect sanitation through collection and removal of human and industrial waterborne waste (sewage) by the construction of drainage channels.

During the first six months, while the legal challenges were in process, the board did meet and accomplished such organizing tasks as electing officers, adopting rules and a corporate seal, establishing a schedule of regular meetings, and appointing officers. However, it was decided that until the challenges were resolved, no funds would be expended or borrowed. During this period, a quorum rarely made it to the meetings of the board.

SDC Board Gets Organized

On June 18, 1890, the first actions were taken in accepting a line of credit from four banks and directing Chief Engineer Lyman E. Cooley to undertake detailed investigations and submit to the board not less than four routes for a channel between the South Branch and Summit or thereabouts. Beyond Summit, the chief engineer was directed to make investigations but was not specifically asked for potential routes.

The work toward construction of the Main Channel to save Chicago was underway.

References

Chicago Sunday Tribune, October 13, 1889.

Marshall J. Wilson v. Board of Trustees of the SDC et al., Supreme Court of Illinois, 133 Ill. 443; 27 NE 203, June 12, 1890.

People v. Murray Nelson et al., The (members of the SDC Board), Supreme Court of Illinois, 133 Ill. 565; 27 NE 217, June 12, 1890.

Rosenberg, Michael. "Why the MWRD Is a Special District" (unpublished manuscript). MWRD R&D Department Seminar, 1997.

SDC. Proceedings of the Board of Trustees of the SDC. 1890.

CHAPTER 3: FORMATION OF THE SANITARY DISTRICT OF CHICAGO

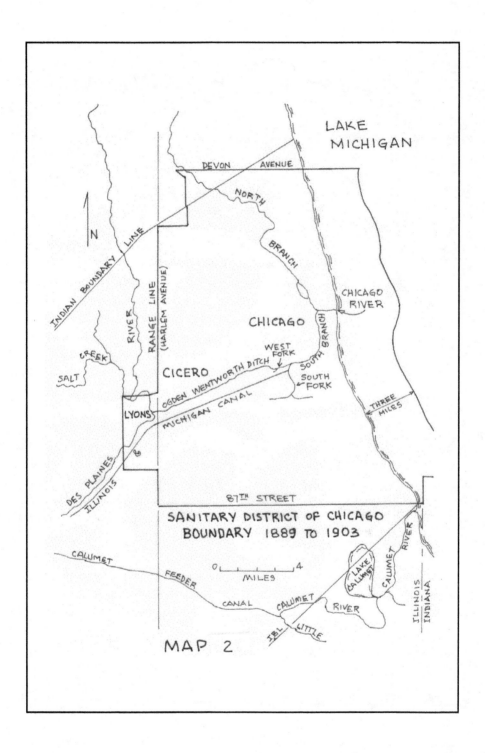

MAP 2

Chapter 4

Finding Direction

In describing the undertaking of this great work, the events of 1890 and 1891 will follow chronological order as during these two years there was not a commonality of purpose among the members of the board. A majority of the board did not seem willing to follow the intent of the act. With three new members in late 1891, a new majority resolved to proceed with a firm grasp of the mission at hand. Thereafter, the work of constructing the Main Channel and reversing the flow in the Chicago River is explained in several chapters, each covering a separate phase of the work. Work proceeded first on the Rock Section from Willow Springs to Lockport, then on the Earth and Rock Section between Summit and Willow Springs, followed by the Earth Section from Robey Street (now Damen Avenue) to Summit. Two phases were undertaken later: the Joliet Project to improve the Des Plaines River from Lockport through the city of Joliet, and the Chicago River Improvement from Lake Street to Robey Street (see map 3).

Getting Started

Due to the legal challenges brought against the SDC, little was accomplished in the first six months. The first specific step was taken in June 1890 in directing Chief Engineer Lyman E. Cooley to undertake work to present the board of trustees with not less than four routes of a channel between the South Branch and Summit and to make investigations of the Des Plaines River Valley. The intent was to build a channel over these 8 miles and to build two pumping stations, one to lift the waters of the South Branch into the channel, and the other to lift the water out of the channel into the Des Plaines River. It was also anticipated that the Des Plaines River would have to be enlarged to accept additional flow.

The board was not all of one mind to build a channel all the way to Joliet. There was divided opinion on the extent of the channel in spite of specific criteria in the act and the 1887 recommendations of the Commission on Drainage and Water Supply. A majority of the SDC Board during the first two years consistently sought a minimal approach, the motive probably being to minimize taxes. The board even considered the use of the I&M Canal instead of building a new channel, something which the commission had suggested as an interim project.

The Chief Engineer Is at Odds with the Board

Chief Engineer Cooley reported progress on this work in September 1890 and proposed a study of the pollution of the South Fork and the amounts of dilution required to put the method of dilution on a sound basis. Cooley had this to say as part of his report:

> A work of this magnitude demands the most careful and thorough work and all work hitherto undertaken or projected is upon the most comprehensive scale and by the most exact methods. It is proposed to leave nothing to guess work or assumption. The margin of certainty to be reached and the savings which may follow such a course ranges in the millions.

In October 1890, the chief engineer reported to the board that he was unable to deliver a long-sought report on relations with the federal government due to preoccupation with engineering work and the board's Committee on Engineering. The following month, the board was micromanaging the chief engineer as he was directed to suspend study of the upper Des Plaines River and to abstain from obtaining information on the cost of land and land ownership records. However, survey work on the Des Plaines and Illinois Rivers downstream of Joliet was allowed to continue.

Cooley was not progressing to the satisfaction of a majority of the board as his services were terminated in December by a vote of five to two, without a mention of the cause. However, President Prendergast mentioned in his annual message in January 1891 that the board received excuses and delays from the chief engineer, and he predicted progress under a new chief engineer. It should be noted that Prendergast was elected president by the board at the meeting prior to Cooley's termination. The president's message did not pinpoint the specific works that were contemplated, but Prendergast speculated that construction would begin in 1891. He also held hope that the federal government would participate in funding the construction of the channel, noting that the Calumet River has been improved with federal expenditures. Time would prove this hope of federal assistance to be unsatisfied.

The Board Picks a New Chief

One of the products of excavating a large and deep channel would be copious amounts of spoil, and the board considered transportation to and deposition in the lake. In an opinion rendered by the board's attorney in February 1891, caution was expressed that the SDC was not authorized to purchase riparian property for the purpose of creating a landfill, and the lake bottom was not available for sale by the state for creating an island. Further, the federal government had clear authority governing any encroachments in navigable waters.

The services of General John Newton were retained as a consulting engineer to the board, and a new chief engineer, William Worthen, was appointed, both in December 1890. Newton, Worthen, and the board began to focus on the cost of works to accomplish the purpose of the act. The cost of a channel from Summit to Joliet and the deepening of the existing canal in the city was estimated and reported in March 1891 to be at $25,900,000. The board determined that that would exceed what could be raised through taxes and the sale of bonds, about $15,000,000. Moreover, the majority of the cost and construction time would be consumed by the need for a large channel cut through rock to meet the requirements of the act. The channel upstream of Summit, to be excavated in glacial till, would be easier by comparison. The board didn't foresee the many problems that would delay this "easy" section.

With this information, wrong as it would eventually prove to be, the board resolved to take action. The board directed that (1) the Main Channel route between the South Branch and the Cook County line be in and along the I&M Canal and then continue to Joliet, (2) they acquire the needed land rights, (3) they prepare the plans and specifications, and (4) they meet with the I&M Canal Commissioners. The board considered relief through amendment of the statute, seeking to allow a channel of smaller yet navigable dimension and to authorize use of the I&M Canal outside of Cook County. Their efforts to amend were eventually rebuffed in Springfield.

Another New Chief

A new chief engineer, Samuel G. Artingstall, was appointed in May following Worthen's resignation due to illness in April 1891. Newton also resigned as consulting engineer. Artingstall was one of the consulting engineers on the Commission on Drainage and Water Supply in 1886 and 1887 and was the former city engineer for Chicago. He came to the job with knowledge of what needed to be done. At the same meeting where Worthen resigned, Frank Wenter, one of the board members who did not support Cooley's discharge, submitted a memo to the board challenging the March 1891 report and estimate of $25,900,000. Although the board voted to file away his memo, it was not forgotten.

In May, Chief Engineer Artingstall reported on four routes of the main drainage channel between the South Branch and Summit. Note here the use of the term "main drainage channel," which the new chief engineer favored and which became the name of choice for the great undertaking. The alternative routes for the channel to Summit were (1) from the West Fork at Western Avenue, following the Ogden-Wentworth Ditch; (2) from the South Fork at Ashland Avenue, following the route of the I&M Canal; (3) from the West Arm of the South Fork near Western Avenue, west along Thirty-Ninth Street to the

I&M Canal, then along the I&M Canal; and (4) following Route 3 to the I&M Canal, crossing the I&M Canal and continuing northwesterly to the Ogden-Wentworth Ditch, then following the ditch (see map 4).

Routes 3 and 4 were favored because they would connect where the putrid influence of the Stockyards was at its worst. To provide relief for the Stockyards pollution in Routes 1 and 2, a relief channel connecting the West Arm of the South Fork to the channel was included in the estimates. The estimated costs, not including land rights, ranged from $2,100,000 to $3,400,000. The report was referred to the Committee on Engineering for consideration, which became a common practice for board decision making. In July, the board selected Route 4 but took no further action toward land purchase or construction.

In his report, Artingstall, without naming Cooley, recognized the valuable and reliable information and data assembled previously, without which the important engineering work of the SDC could not proceed.

New Plan for the Main Channel

Artingstall had greater experience with canal construction methods and costs and reviewed the Newton-Worthen report and estimate of March. He also benefitted from his previous experience with the commission. Artingstall submitted a report to the board in June 1891 proposing a least-cost route from Summit to Joliet, estimated at $14,500,000, reducing the previous estimate by $8,200,000. The least-cost route minimized rock excavation. Artingstall recommended that work proceed initially on the Rock Section from Willow Springs to Lockport since this would be the most expensive, difficult, and time-consuming part of the channel construction. The report, as usual, was referred to the Committee on Engineering.

The reduced cost improved the prospect for financing, and the board directed at the following meeting that the ROW be defined for the route from Summit to Joliet, that an ordinance for acquisition of the land be prepared and submitted, and that design and specification work proceed. In July, the board directed that steps be taken to raise $5,000,000 for land purchase and construction expenditures anticipated over the next eighteen months. In August, the ROW ordinance for Route 4 from the South Fork to Summit was adopted, and the following month, the ROW ordinance for least-cost route from Summit to Joliet was adopted. Purchase of the necessary land rights was underway.

In October, rising out of concern for unsanitary conditions in the Chicago River and negative reports in the press that the SDC was stalling in its work, President Prendergast urged the board to proceed with enlarging and deepening the I&M Canal from Bridgeport to Corwith, a railroad junction near Thirty-Ninth Street and the I&M Canal (about 3600 West) where Route 4 crossed the I&M Canal, and to build a pumping station at Corwith for the I&M Canal to replace the inadequate Bridgeport Pumping Station, allowing its abandonment. Prendergast believed that this work would eventually be needed when Route 4 and the remainder of the channel was constructed. By doing it now, the city would be relieved of the expense of the inadequate pumping works. There was apparently no opposition to this suggestion as the board adopted the same by a vote of six to zero. It was to be sent to the city and I&M Canal Commissioners to initiate formal discussions.

As to the vote, by this time there were only six board members. Three had resigned (King, Nelson, and Willing) during the previous summer of 1891. That there was no debate on Prendergast's proposal may have been indicative that the old majority was losing its grasp. The vacancies resulted in candidates having to run in the November general election to fill the unexpired terms. Elected on November 3 were Boldenweck, Cooley, and Eckhart. Yes, none other than Lyman E. Cooley, the SDC comeback kid, and Eckhart, one of the Committee of Five. All three had the same perspective of the mission before the SDC. It should come as no surprise that in the following month, when the board elected new officers, Prendergast was turned out. To the office of president, Frank Wenter was elected, and he, in turn, designated Cooley as chairman of the Committee on Engineering of the board. Remember that Wenter was one of those who voted *nay* on the termination of Cooley as chief engineer one year earlier. These two would prove to be the dynamic duo in revitalizing the work of the SDC Board.

New Members on the Board

Prior to the turn in officers, the board took action on two items. The construction of the channel would result in the need to build bridges for railroads and roadways to pass over the channel. To conform to the requirements in the act for the channel to be navigable, the board resolved that all bridges will be of the draw or swing type to allow for the passage of boats. On the proposal for improving the I&M Canal and building a pumping station at Corwith, the board referred the matter to the Committee on Engineering.

Artingstall also reported to the board his opinions on methods of excavation for the channel as follows:

- Dredging in water could be done cheaper and with greater facility than by excavation machinery working on dry ground.
- It was impractical to believe that the channel excavation next to the Des Plaines River could be worked in a dry condition.
- Due to the difficulty of rock excavation, it would be more expeditious to begin work at multiple points along the route simultaneously rather than work from both ends toward the middle.
- It is the board's discretion to do the work by contract or by labor hired directly by the SDC.

Actual experience over time did not prove Artingstall's opinions 1 and 2 to be valid, and he apparently overlooked the act's requirement for competitively bid contracts with respect to opinion 4.

President Wenter's annual message in December set the tone for 1892. He called for speedy completion of the work to build the channel, advised the adherence to the act's size and capacity requirements for the channel, declared the first priority to be the Rock Section, predicted that construction would begin by next July, and expressed confidence in the supply of financial resources to complete the work. Following the message, at the urging of Cooley, the board directed the Committee on Engineering to review past work and make recommendations to expedite the commencement of construction of the channel at the earliest date and with the greatest economy.

Review of Past Planning

The Committee on Engineering proceeded with a comprehensive review, including input from experts outside the SDC and from the railroads. The latter took issue with some of the routes selected. The report of the committee, given in January 1892, did not receive unanimous support and was adopted with a six to one vote with two abstentions. The report was critical of the first two years of activity, noting the lack of guidance by "a definite policy or a specific plan of operations."

As a result of the review and input, the committee found the following:

1. The act's requirement of a navigable channel with swing bridges that could be opened was not made known to the railroads. There were ten railroad companies and eight crossing locations. Only fixed bridges were contemplated and estimated.

2. The required capacity of 600,000 cfm in the rock cut was contemplated and estimated to begin at Sag, whereas it should have begun at Willow Springs. (Note: It would later be found through subsurface explorations that rock would also be encountered near Summit.)

3. Engineering data and records were incomplete or nonexistent regarding channel dimensions and grade. It cannot be proven that the construction estimates were based on a channel design that meets the requirements of Section 23 of the act.

4. Every engineer advising the board had recommended that work should begin on the rock cut as the completion of this section would require the longest time. Although the board believed that purchase of this land would take over one year, further examination revealed that it could be acquired in sixty days.

5. The board was preoccupied with work on the Earth Section between the South Branch and Summit, a reach that would be unusable until the remainder was completed.

The contemplated pumping stations would not meet the requirements of the act or provide sufficient relief to the Chicago River.

6. The Earth Section was planned to be connected to the South Fork near the Stockyards, have fixed bridges, be dependent on pumping works, and be completed by 1892 or 1893. No planning or investment was to be made on the remainder of the work beyond Summit.

Cooley concluded the Committee on Engineering report as follows:

> This law was matured after long consideration and is explicit in its provisions. It lays down the conditions which must be met, and definitely prescribes the limitations upon capacity and size of channel. It is no part of the duty of this Board to question these provisions, and it has no option other than to carry them out in accordance with their full spirit.

Strategy for the Future

The adopted report made the following specific recommendations:

1. Reconsider the route from Sag to Lockport and prepare plans for construction in the dry by March; purchase land and begin construction by June.

2. Reconsider the route from Willow Springs to Sag as above, planning to begin construction by September.

3. Reconsider the route from the South Branch to Willow Springs based on the need for adequate capacity and future expansion and plan to begin construction in 1893.

4. Reconsider the route from Lockport to Joliet based on the need for adequate capacity through Joliet, the economy of water power development, the need for navigation, and the possibility of federal cooperation. Construction need not begin until 1895.

5. Fix the minimum flow capacity of the Chicago River so that improvements thereto will conform to a general plan.

Here we have a new majority, taking the board in a new direction twenty-four months after the first meeting of the board and nineteen months after the beginning of actual deliberations following the resolution of the legal challenges. Actually, the direction was not new but was, as Cooley put it, "to carry out [the requirements of the act] in accordance with their full spirit."

Another Change in Chief Engineer

In January, Artingstall resigned and Benezette Williams was appointed chief engineer. Williams was the other consulting engineer on the Commission on Drainage and Water Supply.

Reference

SDC. Proceedings of the Board of Trustees of the SDC. 1890, 1891, and 1892.

CHAPTER 4: FINDING DIRECTION

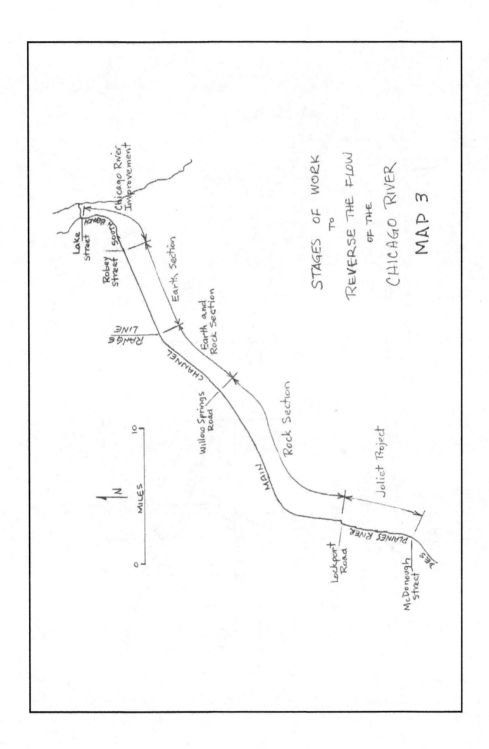

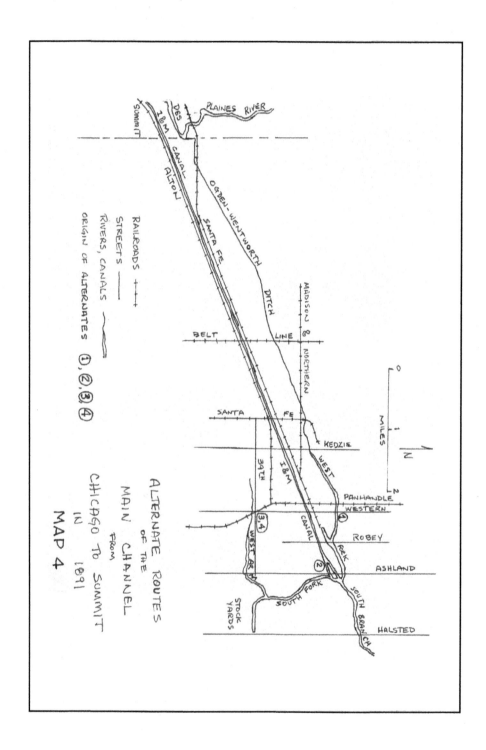

CHAPTER 4: FINDING DIRECTION

Photograph 4.1 taken on July 25, 1894 shows a meeting of the Board of Trustees of the SDC in their offices in the Rialto Building. The Board met twice per month doing most of their deliberations in several committees. By the time of this photograph, contacts for the excavation of the Main Channel had been awarded. Contracts for bridges and other structures were still ahead. MWRD photo, disc 26, image 27.

Photograph 4.2 taken in 1895 shows Stephen Street in Lemont. Northwest of Lemont, this street becomes Lemont Road. The Santa Fe Railroad tracks cross the view in the foreground. Although work had begun on excavation of the Main Channel in Section 8 behind the camera, work was not yet underway to build a new road and railroad bridge and raise the railroad approach embankment. A viaduct would also be built for Stephen Street to pass under the railroad. In addition to Chicago, Joliet and Lockport, Lemont was the only other town of significance along the channel route. MWRD photo, Geiger set, image 301.

Chapter 5

Rock Section

The Rock Section portion of the Main Channel begins at Willow Springs Road and ends at Lockport. Niagaran dolomite limestone is very close to or at the surface of the ground throughout much of this reach from upstream of Lemont to Lockport. From Willow Springs to upstream of Lemont, the rock surface is high enough to require a channel depth of at least 18 feet as required by the act. The board was advised that excavation in rock would be more difficult and time consuming than excavation in earth. For this reason it was decided that construction of the Rock Section should proceed first as it would be the most critical part of the completion of the Main Channel and the reversal of the Chicago River. This proved not to be the case as other difficulties intervened to delay the overall completion of the work (see map 5).

Introduction

Over the first half of 1892, the focus of work by the SDC was the selection of a route between Willow Springs and Joliet and the preparation of channel design and specifications. This reach was of primary concern because of the thin soil cover and the extensive amount of rock excavation required. Because of the extensive amount of rock, it was initially thought that the Rock Section would take the most time to construct.

The problem downstream of Lockport referred to by Cooley was the controversy over whether to develop water power at the end of the channel and make the channel navigable all the way to Joliet. There was also a difference of opinion on the method of disposal of the excavated rock: whether to spoil the excavated material along the channel route in a huge windrow or remove the spoil to allow industrial development along the channel frontage. Since cost was an overriding issue, it was decided to let the contractors bid on alternatives. Then the board could select the alternative that was within their means.

In this context, *water power* did not necessarily mean hydroelectric generation and transmission, which was eventually implemented in 1908. Water power also referred to mechanical power transmission, such as a water mill. High-voltage transmission of electrical current was not yet perfected for commercial use. Thus, if electrical generation was being considered, it would out of necessity be at the location of use, such as an industry in Joliet located close to the generating station. Transmission of electricity back to Chicago would not have been economically feasible in 1892. The first transmission of high-voltage, three-phase electricity was installed in 1895 in Folsum, California. Nikola Tesla discovered alternating current in the late 1800s, but it was not perfected for transmission until the 1890s. Electricity generated at Niagara Falls, New York, in the early 1890s was used locally, and it was not until November 1896 that electricity was transmitted to Buffalo for use.

The selected route between Willow Springs and Joliet would follow the low part of the Des Plaines River valley to minimize the depth of rock excavation. Since the I&M Canal followed the south or east side of the valley, the channel route would lie north or west of the I&M Canal and south or east of the river. Where the river meandered to the south or east side of the valley floor, these meanders would be cut off by construction of a river diversion channel and a levee built to separate the river and channel (see map 5).

Bidding on Alternative Channel Schemes

As the land along the route was being purchased, bids for construction were advertised in April 1892. The specifications divided the work of the 17.5-mile channel into several contracts called sections. A total of seventeen sections were contemplated. The specifications also called for bids on three propositions, or alternatives. All three propositions included the excavation of the Main Channel in rock from Willow Springs to Lockport and the construction of channel walls where necessary. This channel was to be 160 feet wide and of sufficient depth to meet the statutory requirements. From Lockport to Joliet, the rock surface dropped in elevation such that the channel would be above grade and would have to be confined within embankments.

Proposition 1 included the construction of a railroad to haul excavated material from the sections upstream of Lockport to the embankments downstream of Lockport. The contractor in each section would install the tracks, but the SDC would purchase the railroad equipment and operate the trains to haul the spoil. This proposition also included a dam and water power at Joliet.

Proposition 2 included only the channel excavation between Willow Springs and Lockport, requiring fourteen sections rather than eighteen, and spoiling all material along the route on additional ROW purchased by the SDC. This proposition set aside the need for a channel from Lockport to Joliet, recognizing concern by several board members regarding the feasibility and cost of water power development. Proposition 3 was the same as Proposition 1, except the purchase of railroad equipment and operation of the trains was left to the contractors.

Bids were opened in June 1892, and Chief Engineer Williams was asked to analyze the bids and prepare a recommendation. Based on the low bids only, the cost of Propositions 1, 2, and 3 were $10,192,800, $10,696,800, and $17,105,900, respectively. In his analysis, Williams included the estimated cost of land, railroad operation, alternative channel improvement downstream of Lockport for Proposition 2, as well as supervision and other incidentals in all three propositions so that all were on the same basis. The resulting total cost of Propositions 1, 2, and 3 were $13,667,000, $12,815,900, and $18,091,100, respectively. Williams also promoted the inclusion of water-power development supported with an estimate of revenues provided by Westinghouse Electric Co., a manufacturer of hydroelectric generators. Williams included testimonials of the benefits derived from a water power plant in Marseilles, Illinois, and visions of industrial development along the channel frontage. He concluded by recommending Proposition 1 for implementation.

The matter was under study by a board committee. In the committee report, little attention was given to some of the estimates in Williams's analysis, doubt was expressed over the potential for industrial development in the near term along the channel frontage, reluctance was expressed for taking on the burden of ownership and operation of a railroad, and as a priority, water power development was far below sanitation and navigation. The committee concluded by recommending that Proposition 2 be implemented.

The issues of water power and a navigable channel to Joliet were to be set aside for the present. They were not pursued for another ten or more years as part of the extension of the Main Channel. However, the work necessary to accommodate the additional discharge of water from the Main Channel to Joliet was taken up in 1893 in what became known as the Joliet Project, described in chapter 8.

Award of the First Contracts

The full board concurred in the committee recommendation, and fourteen contracts were awarded in June 1892 to six different contractors, dividing the 74,000 feet (14 miles) of channel into fourteen sections. The total bid cost for the channel excavation and retaining walls was just under $13,000,000. Seven of the sections were 5,000 feet in length, and the others varied from 4,700 to 6,000 feet in length. The fourteen sections extended from near Willow Springs Road to about 1 mile north of Ninth Street in Lockport. All the contracts were executed in July, and the contractors were ordered to proceed with the work by September 1. The contracts called for completion of work by 1896. The first breaking of ground actually occurred on August 24, 1892, prior to the groundbreaking ceremony, as explained later on September 3.

The sections were numbered from north to south. Section 1 began just downstream of Willow Springs Road. Sections 2 through 7 spanned the distance almost to Lemont, and there were no crossings, only a few small bridges that led to the I&M Canal. Section 8 at Lemont included the Lemont Road (Stephens Street) and Santa Fe Railroad crossings. Section 10 had a private road crossing leading to a quarry. Section 12 included the Romeoville Road crossing, and Sections 9, 11, 13, and 14 had no crossings (see map 5).

Shortly after the start of work, one of the contractors complained that the location of the channel as staked out on the ground by SDC surveyors was materially different from that shown on the plans. The board referred the matter to Chief Engineer Williams, and he met with the contractors and reported back to the board that there had been numerous changes made in the field as is normal field engineering practice. Only two contractors found the changes unacceptable. The board referred these two complaints to the attorney and engineer for a recommendation, but no further action was taken. The contractors dropped their objections so that they could proceed with the work.

There was a change made in February 1894 to the beginning limit of Section 1. After award of Section A, which is upstream of Section 1 in the Earth and Rock Section, it was found that several hundred feet of retaining walls were required but that insufficient rock existed in Section A to build the walls. The board acted on the recommendation of Chief Engineer Randolph that since ample rock existed in Section 1, Section A be shortened by 700 feet and Section 1 lengthened by a like amount. This placed the Willow Springs Road crossing and the transition from a rectangular channel prism to a trapezoidal prism in Section 1. A more detailed explanation of this change is found in chapter 6 (see map 6).

Contract Requirements

All fourteen contracts, actually called agreements, were identical. The contracts were divided into clauses each sequentially identified by a letter and, except for the first three clauses, were also identified with a topic. Clause A set forth the basis of the agreement and the responsibilities of each party and cited applicable statutory authority. Clause B declared that the work was under the supervision of the chief engineer. Clause C defined the *SDC, engineer*, and *contractor* as used in the agreement. Clause D included the technical specifications and was further divided by topics sequentially identified by number.

Clause D, Specifications, included the technical details and requirements in twenty-four items. Item 1, Location, gave the alignment for the Main Drainage Channel. The line of the channel was referenced to the I&M Canal. The channel varied from 460 to 730 feet from the straight segments of the I&M Canal. The alignment was also shown on plan 2a, made a part of the contract documents. Also referenced and made a part was a report of the board's Joint Committee on Engineering and Finance (JCEF), which modified the alignment by making reference to horizontal clearances from the Santa Fe Railroad tracks, both upstream and downstream of Lemont. Downstream of Lemont, the tracks were to the east of the channel and as close as 50 feet near Romeoville Road. Upstream of Lemont, the tracks were to the west of the channel, and the horizontal clearance of 800 feet provided for the Des Plaines River diversion channel and levee between the tracks and the channel (see map 7).

The limits of each section were defined in Item 2, Sections, by 100-foot stations along the centerline of the channel. It also referenced Sections 1, 2, 7, 8, 9, and 10 as requiring river-diversion work; Section 9 as possibly requiring relocation of a portion of the I&M Canal; and Sections 8 through 14 as possibly requiring Santa Fe Railroad track changes. Item 3, Grade, specified the grade of the channel bottom as 24 feet below datum with a slope of 0.08 feet per 1,000 feet, referencing the 1847 I&M Canal datum. Allowance was included for the SDC to change the grade.

Item 4, Dimensions of Cross Sections, went into great detail on construction techniques. The sides of the vertical rock cut, or channel prism, were to be 160 feet apart at the base with 6-inch horizontal setbacks to accommodate the specified use of channeling machines. These machines cut a vertical slot to define each face of the channel prism to prevent the fracturing of the remaining walls when dynamite was used to fracture the rock within the channel prism for removal. The walls of the channel were "to be left as smooth and solid as can be obtained with a skillful and proper use of a channeling machine." Setbacks were allowed at vertical intervals not less than 16 feet. Thus, throughout most of the channel in full rock cut, the channel prism is 162 feet wide at the top of the cut.

Item 5, Retaining Walls, specified the construction of structural walls, to be installed where the top of rock is below a level 5 feet above datum. The walls were to be built of dry rubble from limestone blocks salvaged from the blasting. Details were given on construction technique, minimum and maximum block size, and wall foundations. The channel face was to be slightly sloped back so that the top of the wall is 83 feet from the channel centerline. The back of the wall was stepped in offsets and typical wall cross sections were provided. Wall thickness at the top was a minimum of 4 feet, and the thickness at the bottom was a minimum of one half the height.

Possible changes in the I&M Canal in Section 9 were provided for in Item 6, Canal Deviation. Item 7, River Diversion, specified the river diversion channel in the pertinent sections, the east or south side of which was to be no closer than 360 feet from the centerline of the Main Channel. The width of the diversion channel was 200 feet with side slopes of 1 on 1. The spoil was to be placed between the diversion channel and the Main Channel. Item 8, Clearing and Grubbing, dealt with clearing the land surface of the ROW.

Item 9, Levee, required the contractor to build levees to protect the work from flooding. Item 10, Railway Changes, provided that the contractor may be required to change the location of railway grades. Detailed requirements of embankments were given. Item 11, Disposition of Material, required the disposal of all excavated material along the channel route on the ROW provided. The toe of the spoil pile was to be no closer than 50 feet from the channel. Item 12, Drainage, required the contractor to provide for pumping all drainage from their section to the Des Plaines River without interfering with other contractors.

One of the contentious parts of the contract would prove to be Item 13, Classification of Material, in which material to be excavated was classified as either *glacial drift* or *solid rock*. Glacial drift included everything in whatever condition that overlaid the bedrock. Solid rock was all rock material found in its natural condition, even if it was loosened and could be removed without blasting. Item 14, Quantity of Material, referred to a tabulation of the quantities of glacial drift and solid rock in each section and to a profile of the ground and rock surface along the channel, which were part of the contract document. The contractor was to assume all risks regarding variations with the quantities of materials to be excavated.

Item 15, Highways, required that all highways and roads be kept open and unobstructed. This requirement was applied to railroads and canals as well. In Item 16, Bridges and Structures, the SDC reserved the right to enter the contractors' work area to build bridges, roads, or other structures but not to interfere with the work of the contractor. This requirement also provided that such work could be done by the contractor, at the expense of the SDC. Item 17, Explosives, governed the use and storage of explosives. Blasting materials were to be stored in bulletproof buildings, not to be stored in quantities greater than 5,000 pounds, and to be used only by skilled and careful workmen.

Item 18, Measurement, established cubic yards for the measurement of quantities. For payment purposes, quantities were to be determined from the survey notes of the engineer. Payment was based on the quantities of glacial drift, solid rock, and retaining wall volume. Work that exceeded the dimensions specified would not be paid for. All work under the contract was to be compensated for through the unit prices for these three measured quantities. Item 19, Extra Work, provided that extra work could only be approved by the engineer and could only be compensated for at the unit prices for one or more of the three measured quantities. However, the engineer could approve another more reasonable method of compensation.

In Item 20, Under Responsibility of Contractor, the contractor was responsible for the entire work. All work was subject to inspection and approval by the engineer, and failure to comply with the contract would lead to forfeiture. In Item 21, Changes in Plan, the SDC reserved the right to make changes. The contractor was obligated by Item 22, Tools, to furnish all tools required for the work and to remove same upon completion. The contractor was required in Item 23, Precautions, to take necessary precautions to protect the work and to avoid accidents and incidents which might delay the work. Item 24, Workmen, required the contractor to supply all workmen necessary to complete the work and provide competent supervision and experts where skills are required.

Damages, Clause E, that were caused by the contractor or persons in their employ were the responsibility of the contractor. The SDC had the right to have the engineer estimate the damages and pay same to the damaged party and deduct a like amount from the contractor's payment. The SDC also had the right to withhold payment from the contractor for claims made by third parties or subcontractors for services performed for work under the contract. Clause F referred to the Act of 1889 as authority for the work and specifically pointed to the requirement for contractor employees to be citizens or intending to become citizens.

Time, Clause G, required the contractor to begin work within thirty days of contract execution, subject to availability of ROW and an order to proceed from the SDC. Work was to have been completed by April 30, 1896, or forty-five months following the order to proceed. The work was to progress in proportion to the time elapsed, the first four and last two months not being included in the proration. The contractor may proceed more rapidly, but the SDC was not obligated to pay more than the prorated progress would allow. Prices, Clause H, set forth the unit cost for glacial drift, solid rock, and dry-rubble masonry. Full compensation for all work performed under the contract was based on these unit costs.

Time and Manner of Payment, Clause I, provided for progress payments on the tenth and twenty-fifth of each month based on certification by the engineer of work performed, with 12.5 percent retainage until completion and acceptance of all work. Grading of Prices, Clause J, allowed the SDC to further reduce any progress payment by up to 10 percent, if the engineer determines that based on any disparity in unit costs, the contractor has not incurred actual costs commensurate with the work performed to date. Liquidated Damages, Clause K, provided that the SDC could keep the 12.5 percent retainage as liquidated damages if the contractor failed to prosecute or complete the work as specified.

Certificate, Clause L, allowed for the payment of the retainage at the completion of the work upon receipt of a certificate from the chief engineer indicating that all work had been completed. Failure to Complete, Clause M, allowed the SDC to complete the work of the contractor either with another contractor or by its own forces upon certification by the engineer that the work is abandoned, is assigned to others by the contractor, and that control of the work has been lost or if delay will not allow completion by the completion date. In such an event, all monies owed to the contractor could be used by the SDC to pay for the completion of the work. Failure to Pay Laborers, Clause N, allowed the SDC to pay monies owed to employees or suppliers of the contractor for work completed if the contractor failed to make timely payment. Such payments could be deducted from monies owed to the contractor.

One last clause, titled Health Regulations, but not given the letter *O*, required contractors to enforce regulations protecting the care and health of employees by governing the cleanliness of dwellings, supply of potable water, removal of waste, and prevention of nuisances. This and other above provisions caused difficulties in contract administration, resulting in delays and other problems. Some problems caused the SDC to undertake other means to deal with issues that proved to be beyond the reasonable control of the contractors. Problems in contract administration for the Rock Section are described later in this chapter. Other general problems are explained in chapter 11.

Shovel Day: September 3, 1892

A celebration was planned for September 3, Shovel Day, to observe the official breaking of ground. A special train carried the trustees and invited dignitaries to Lemont, whereupon the entourage traveled by a parade of horse-drawn carriages and wagons to the centerline of the channel route on the Cook-Will County line in Section 10. On a platform erected for the occasion, President Wenter and trustees Cooley and Eckhart gave speeches. The turning of a shovel of earth and a blast of dynamite signified the start of construction. Other speakers instrumental in the activities leading up to this day—including congressmen, state assemblymen, and notable locals—also gave speeches. Lavish accolades were bestowed upon those persons instrumental in the accomplishments leading up to the day, and great visions of the future were promised. One speaker, Fernando Jones of Chicago, had witnessed the digging of the I&M Canal half a century earlier. Illinois governor Fifer and Chicago mayor Washburne sent representatives.

Work commenced and was followed with intense interest by the board, with monthly reports starting in the summer of 1893. These reports detailed the quantity of men and equipment on the job sites for each section and reported progress in terms of the volume of excavation. While efforts were devoted to putting work underway in channel excavation upstream of Willow Springs, attention was also given to addressing modifications and problems in the work on the Rock Section.

Change in Grade of the Channel and the Fifth Chief Engineer

In January 1893, Chief Engineer Williams reported to the board that he intended to recommend a change in grade for the Main Channel for the reason of reducing rock excavation. In his March report, he indicated that the channeling machines were producing such a smooth cut that the frictional resistance to the flow of water would be less; thus, the channel gradient could be lowered because the hydraulic gradient would be less. The contracts allowed for a change in grade, so it was appropriate to proceed. By starting at the upstream end of the channel and changing the grade from 0.08 foot of drop per 1,000 feet to 0.053 foot, the channel bottom at Lockport would be 2 feet higher and a savings of 500,000 cubic yards of rock excavation could be realized. The matter was referred to committee for consideration.

The committee referred the matter back to the chief engineer in June, citing the undesirability of the loss in capacity that would result if the recommendation was implemented. The resignation of Chief Engineer Williams was accepted at the board meeting on June 7, 1893, and at the same meeting, the fifth chief engineer, Isham Randolph, was appointed. The board asked that alternatives be investigated. Chief Engineer Randolph reported back in August that, based on the work of his assistant, T. T. Johnston, the reduction in slope starting at Chicago, which would have raised the bottom grade at Lockport as proposed by Williams, was unacceptable. Randolph suggested that by starting at Lockport and lowering the bottom gradient back toward Chicago, the greater depth of the channel and cost of excavation would be offset by reduced costs for the work downstream of Lockport. The extra cost was estimated at $200,000 for the Rock Section alone. The committee agreed, citing several advantages, such as better control and regulation, no loss in capacity with an ice cover, simplification of the work downstream of Lockport, and a more suitable navigable channel. An order was adopted by the board in September calling for a bottom elevation 29.9 feet below CCD at the Main Channel terminus at Lockport, 0.05 foot rise per 1,000 feet to Willow Springs Road, and 0.025 foot rise per 1,000 feet thereafter to Robey Street.

The logic in the above change is not clear, and the advantages cited were never explained or quantified elsewhere. Although not mentioned, Randolph probably doubted Williams's opinion that smoother walls would have compensated for the reduced depth. Nevertheless, this change held and became the basis for construction. Appropriate change orders were issued for the fourteen contracts in the Rock Section.

River Diversion Channel, Goose Lake, and the Levee Height

The SDC was building two channels, not one. The other channel was the Des Plaines River diversion channel, necessary where the natural river channel meandered to the east or south of the valley bottom and was in the way of the Main Channel construction. In addition, a levee was needed between the diversion channel and Main Channel to prevent high water in the Des Plaines River from flooding the excavation of the Main Channel. While diversion channel work was not necessary in all sections, the levee was needed throughout the length because the river was at a higher elevation than the top of the wall of the new Main Channel to a point south of Romeoville Road.

The contract requirements indicated that diversion channel work was required in Sections 1, 2, 7, 8, 9, and 10, but it was also indicated at the time of award that some diversion work was yet to be defined. Chief Engineer Williams reported to the board in January 1893 on the quantities of diversion channel excavation required. It showed that such work was required in Sections 1, 2, 4, and 6 through 11. Later, in April, Williams reported on the line and grade for the diversion channel between Lemont and Romeo. Recommended changes were adopted in July, establishing the width of the top and the grade of the levee for a length of a few miles upstream of Lemont.

Upstream of Romeoville Road on the natural Des Plaines River was Goose Lake. The rock ledges causing this lake were also retarding flood flows, which increased the risk of overtopping the levee. Chief

Engineer Randolph recommended, and the board concurred, that the ledge be removed. In the following month, the grade of the levee top was established throughout the reach from upstream of Summit to Romeoville. The grade of the levee was incorporated into the specifications of all the contracts as follows: from plus 20 feet above datum at Willow Springs Road descending uniformly to 8 feet above datum at Romeoville Road.

Work on the diversion channel and levees continued to draw the interest of the board due to the risk of flooding the construction site of the Main Channel. In November 1893, it was reported that diversion channel work was essentially complete. Contractors were periodically notified to protect their work from flooding. The condition of the levees was reported in detail in December 1894.

Retaining Walls

The requirement for dry-rubble masonry retaining walls assumed that there would be an adequate supply of suitable stone that could be quarried to the appropriate sizes. In dry-rubble masonry, mortar is not used and the stone pieces must fit together tightly. Construction progress indicated that the assumption of masonry stone availability did not prove valid, and the specification could not be reasonably enforced. The board agreed with the chief engineer's recommendation in August 1893 that the specification be relaxed and an alternative of cement masonry walls be studied. Cement masonry allows the use of cement mortar to bind the stone and fill voids in the joints. After study and further observation, Randolph recommended in May 1894 the substitution of cement masonry for dry-rubble masonry and appropriate price adjustment for the walls in several sections. The adjustments were determined, and the contracts amended by agreement for Sections 1, 3, 5, 6, 7, 8, 11, 12, and 13.

The contractor on Section 14 would not accept the adjustment, and eventually, the SDC exercised its right to have the work performed by another contractor. Advertised in December 1895, bids were opened and an award made in February, but the contract was not executed until May 1896. Due to the lack of suitable masonry stone, the contractor was allowed to construct the wall using concrete.

Conglomerate Rock in Glacial Drift

Sections 2, 3, and 4 had several feet of glacial drift over the solid rock, and the same contractor held all three contracts. In excavating the glacial drift, large pieces of conglomerate rock were found. The contractor requested that they be compensated at the price for solid rock for removal of this material as it could not be removed at the price for glacial drift. Chief Engineer Randolph reported his decision to reject the claim. Upon review, the board upheld the decision. The controversy continued because the contractor refused to accept the decision.

The chief engineer was requested to explore alternative methods of removal, and a test was authorized for hydraulic pressure and sluicing of the material. The test proved inconclusive, and there was no movement on the earlier claim decision. The contractor refused to proceed, and eventually their lack of performance led to forfeiture and readvertisement. A settlement was worked out such that the contractor agreed to accept the decision if he could continue on Sections 2 and 4. Allowing the contractor to continue for these two sections was acceptable, and Section 3 was awarded to another contractor. Conglomerate in the glacial till also became an issue in Sections 1 and 5; however, in these sections the matter was resolved by forfeiture. Section 1 was readvertised, and Section 5 was voluntary reassigned to another contractor.

Last Section

One additional section was eventually added south of Section 14. Section 15 was also largely cut out of rock and extended the channel to its initial terminus, about one-third of a mile north of Ninth Street in Lockport. Section 15 was 4,000 feet in length, with the first 1,000 feet being of the typical 160-foot wide rock cut and the remainder being widened gradually over a distance of 3,000 feet to provide a turning basin for watercraft. The need for a regulating control structure was first outlined in Chief Engineer Williams's report in March 1893 on the Main Channel tailrace and work from Lockport to Joliet.

Section 15 is 450 feet wide at the channel terminus, and this provided a location for the discharge control gates. The southern end of this section required walls above the surrounding terrain because the land surface was lower. The Section 15 excavation and retaining wall contract was advertised in June 1894, with bids to be opened in August. The award was made, and the contract executed in September 1894. The delay of two years since the contracts for Sections 1 through 14 were let was due to the efforts to get work underway on the channel upstream of Willow Springs (see map 8).

The contract for Section 15 was similar in many respects to the contracts for Sections 1 through 14, but the differences show what the SDC had learned from the first round of contracts. More details were shown on plans, cement masonry rather than dry rubble walls were required, the levee between the river and channel was given dimension and grade, and payment to laborers was to be made in US currency.

Setting of the Tablet

Some of the contractors made very good progress, and Section 10 was the pacesetter and the first to complete rock excavation. In September 1893, one year after the start of work, the contractor had reached the bottom grade of the channel, a depth of 25 or 30 feet in solid limestone. The chief engineer reported in July 1894 that some contractors were expected to complete their work in 1895, one year earlier than called for in the contracts. In August 1895, the Section 10 contractor had completed all excavation, and this called for a celebration.

In a virtual repeat of Shovel Day three years earlier, a celebration was held at the same location on September 3, 1895. Again an entourage of dignitaries, this time somewhat smaller in number, set out from Chicago by train to the Cook-Will County Line. With speeches and fanfare, a huge polished granite tablet with raised lettering was set into the north wall of the channel on the county line. The tablet measures 4 by 6 by 1.5 feet and weighs about 3 tons. It marks the county line, the start and end dates of Section 10, and carries the name of the SDC. Remarkably, the tablet remains as it was set and is clearly legible today, 116 years later.

President Wenter made reference to the majestic trough cut in rock, the likes of which were never before seen by man. He glorified the ingenuity and perseverance of the contractor. Commenting on the experience of a recent visitor to the work, he forecasted the application of these construction techniques to the cutting of a canal through Nicaragua. With unbounded optimism, he predicted the rush of Lake Michigan water through the cut two years hence. Wenter even held hope for future development along the channel frontage and water power at Joliet. Obviously, Wenter had gone overboard with optimism.

The guest speaker for the celebration, Judge McConnell, waxed long and eloquent on this great undertaking and wound up by reminding the crowd that this channel was a monumental endeavor, exceeding even the size of the Suez, the Manchester, the Baltic, and North Sea channels or any other by the hand of man.

Controlling Works at Lockport

The last two contracts to complete the Rock Section were for the regulating works at the channel terminus. As explained earlier, the need for regulating works was identified by Chief Engineer Williams in March 1893. The work provided for in these contracts was located in the last 1,000 feet of the west wall of Section 15. The Lockport Controlling Works consisted of the 160-foot-long Bear Trap Dam; 7 vertical sluice gates, each being 30 feet wide; 8 sluice gate bays that were bulkheaded for future use; and foundations, dam end walls, gate piers, and gate house. The Bear Trap Dam is similar to a sector gate, allowing water to pass over its crest and being raised or lowered to control the rate of water passage. The sluice gates are vertical slide gates that allow the passage of water under the gate. These gates are 20 feet high, with the top of the gate 5 feet above datum, the same as the top of the channel walls.

Although it appeared odd that the SDC separated the work into two contracts, and the timing of the two appeared puzzling, there was good reason. The first contract was for the fabrication and installation of the Bear Trap Dam, the 7 sluice gates, the sluice gate hoisting equipment, the construction of the foundation and masonry piers for the 7 sluice gates and the 8 unused gate bays and a small building, wagon bridges connecting the 16 gate piers, and a catwalk connecting the tops of the 16 piers. The second contract, awarded one year later than the first, was for the foundation and end walls for the Bear Trap Dam, the bulkheads in the eight unused gate bays, and other miscellaneous work. The first contractor was responsible for the design of the Bear Trap Dam; therefore, the second contract could not be prepared until the details of the design were known and approved by the SDC. Another reason for the apparent relaxed schedule for these two contracts was that the overall work for channel and bridge construction was behind schedule, and the regulating works could not be tested and used until all other work was complete and there was a full depth of water in the channel.

Chief Engineer Randolph reported to the board in early October 1894 that the design of the regulating works was well underway and that he wished to inspect other similar works in the Midwest. In the company of his assistant, T. T. Johnston, several works were inspected along the Ohio and Kanawha Rivers between Louisville and Pittsburgh. Following that, inspections of structures were made in Minneapolis on the Mississippi River, nearby on the Saint Croix River, and the lock and dam near Sault Ste. Marie. Later that same month, Randolph reported to the board on his inspections, indicating that the present design was as advanced as any they had seen but that some of what they observed could be incorporated into the design. At the Nevers Dam on the Saint Croix River, they were greatly impressed with three Bear Trap Dams, the longest of which was 80 feet, and the ease with which the dams could be operated by one person. A model was built to demonstrate this device to the board members.

In August 1895, the chief engineer reported that plans and specifications were completed and ready for advertisement. He recommended that the work be placed under contract soon since there was a large amount of ironwork involved. The masonry could be worked on the following spring. Advertisement was authorized in September, with bids opened and a recommendation for award made in December. Award was made in January 1896 and amended later that same month with some changes in the design, modifications in general conditions to give the engineer greater control over the work and increasing the responsibility of the contractor, extending the completion from January 1 to May 1, 1897, and changing method of payment from progress payments to payments based on milestones. The last milestone included an operational test of the gates, the first such test in the first contract for the SDC involving operating machinery.

This contract was significantly different than the previous contracts for channel excavation and retaining wall construction. The technical specifications were separated from the general conditions, and the specifications went into great detail regarding construction and manufacturing technique, materials testing, and workmanship. The general conditions included requirements that the contractor was responsible for all liability and costs for patents. Some prices were based on lump sum cost. The relationship and responsibility of the work of the other contractor for the regulating works was spelled out. As a result of the material specifications and testing requirements, the SDC engaged its own testing laboratory service.

The second contract proceeded along a more tortuous path to award. The chief engineer delivered plans and specifications in July 1896 for advertisement. In September, only one bid was received, and it was returned. Readvertised in February 1897, a good set of bids was opened in April and an award was approved. The contract was not executed until June, however, due to some modifications and the securing of a bond by the contractor. This contract is similar to the other contract for the regulating works. Emphasis was placed on completion of the Bear Trap Dam foundation so that the first contractor could proceed with installation of the dam.

Work proceeded without interruption on both of these contracts, although there was a dispute on a claim. Final certificates were issued in July and August 1899. The vertical gates were lifted by manual hoists. At some later time, a steam engine was added to power the hoisting machinery. The Bear Trap Dam was operated with water in tanks and the adjustment of buoyancy and counterbalance weights. The Bear Trap Dam, being designed to be lowered to allow water to pass, provided a means to flush floating debris on downstream. This type of gate also allowed for better control in maintaining a constant flow in the face of a fluctuating lake level at the other end of what would eventually be a 34-mile-long channel open to Lake Michigan.

Bridges

The 14 miles of the Rock Section traversed a region still relatively remote in the 1890s. The Chicago and Alton Railway and Santa Fe Railroad followed the Des Plaines River to Joliet. There were a few small towns or settlements in the river valley. Only three road bridges and one railroad bridge were needed to provide crossings for the Main Channel. Only two road bridges and one railroad bridge were needed to provide crossings of the diversion channel. These were all constructed under separate contracts, with the substructures (foundations) and superstructures often being in separate contracts. The contracts for excavation of the channel assumed that permission to cross the ROW of the road or railroad owners would be obtained without delays being caused. This was not the case of the Santa Fe Railroad crossing in Section 8 as explained later in the discussion of the construction seasons (see map 7).

The excavation contractors were given extras for the construction of temporary bridges so that their work could proceed. The temporary bridges were timber trestles built up from the bottom of the channel on detours around the site of the permanent bridge. Policy as well as technical issues had to be dealt with, which delayed the start of permanent bridge construction. These are discussed in chapter 10. The permanent Willow Springs, Lemont, and Romeoville Roads and the Santa Fe Railroad Bridges over the Main Channel were begun in the summer of 1898 and completed a year later.

Pumping of Drainage

Each contractor was responsible for the pumping of drainage from their work area to keep the area dry. The discharge of this water went to the Des Plaines River. Aside from rainwater, the drainage consisted of groundwater seepage, and as the excavation progressed, the amount of seepage increased. Those contractors who completed their work early were obligated to continue the pumping of drainage for the duration of the contract. Contractors usually left the contract section end wall in place to control drainage. Eventually, these contractors requested to be released from this obligation by the April 1896 completion date. However, the board interpreted the contract language to require the contractors to be obligated until the commission called for in the act could inspect the channel and advise the governor of its completion, provided that this could be accomplished in a reasonable time.

It soon became evident that inspection by the commission would not occur soon, and the obligation for continued pumping by the contractors was viewed as unreasonable. In May 1896 a contract was awarded for a pumping plant in Section 14, which would be maintained and operated by the SDC forces. Thus, all drainage from Sections 1 through 14 would be handled at this location. Furthermore, the board determined that as work was completed in any section, the SDC would assume the responsibility for drainage if the accumulated water in the bottom of the excavation could not be drained by gravity to Section 14. In these situations, the board agreed to purchase or rent and operate the pumping equipment of the contractor.

The pumping plant in Section 14 was not able to maintain a water level at grade, and it was supplemented with a similar plant in Section 15 located on the sill of the sluice gates at the Lockport Controlling Works. This pumping plant was operated until May of 1899, whereupon it was removed for use elsewhere. The pumping plant in Section 14 once again provided the only means of drainage for the Main Channel until December 1899 when it was removed.

Earthquake Damage

Although earthquakes are rare in the Chicago area and are not known to cause significant damage, this phenomenon was but another challenge to the SDC. Construction of the Rock Section portion of the Main Channel was impacted by an earthquake in 1895. On the morning of October 31, 1895, an earthquake shook Chicago and the Des Plaines River valley. A mass of rock was found the next day in Section 7 in the south wall of the Main Channel to have moved slightly toward the channel centerline. This mass had been part of the wall of a former quarry, and the channel retaining wall had been built upon it, closing the gap in the former quarry wall. The earthquake-induced movement resulted in severe cracking and damage to the retaining wall. A 35-foot section of retaining wall was removed and rebuilt as a contract extra.

Nearly a year later in Section 5, close to the present-day Sag Junction, an SDC field engineer noticed signs of failure in the bedrock below the north retaining wall on September 21, 1896. This retaining wall had been constructed prior to the earthquake. A mass of rock 7 feet high from the channel bottom to the top of bedrock sloughed off. The retaining wall above showed no signs of cracking or failure. Two days later, in the night, 250 feet of wall failed, having been pushed into the channel and buried by the mass of muck and soft clay that was behind the wall. The slide extended across the channel bottom to within 30 feet of the opposite wall. It could not be proven that this failure was caused, wholly or in part, by the earthquake. Under extra work, the contractor removed the slide mass and the failed wall, excavated unsound rock, and rebuilt the retaining wall.

Contractor Continuity and Completion of the Work

There were numerous problems with most of the original six contractors in the progress of the work or in disputes. In fact, only Sections 2, 4, and 10 through 13 were completed by the original contractor. Eight different contractors completed the work on the fourteen sections, with one contractor completing six sections and another completing two sections. Disputes over the cost of extras or claims resulted in abandonment and forfeiture on some sections, and in others there was simply a lack of diligence. Sections 1, 3, and 14 were readvertised and awarded to another contractor. Section 14 ended up in court as the original contractor tried to block the award. In several sections, the original contractor agreed to an assignment to another contractor.

Sections 9 through 13 were completed in 1895. Section 1 was completed in 1898 by the second contractor after forfeiture and readvertisement in 1894. Section 8 was completed in 1896, except for several hundred feet of rock, which remained until 1898 when permission from the Santa Fe Railroad was finally obtained. This segment of rock was not completed until late in 1899. The remaining seven of the first fourteen sections were completed in 1896. Section 15 was completed in three years by the original contractor in 1897. The two contracts for the regulating works in Section 15 were completed by the original contractors in the summer of 1899, with some adjustment to the operating mechanism remaining to the end of 1899.

Construction Progress and Methods

The Rock Section was actually composed of two different construction sequences for the various contract sections. Sections 1 through 7 had significant depths of glacial drift overlying the solid rock, whereas Sections 8 through 14 had very little overburden and, typically, solid rock for the full depth of the channel prism. For the latter, the contractors removed whatever overburden existed and began immediately to cut, drill, blast, and remove the rock. For the former, the contractors had to remove the overburden before tackling the rock. Overburden removal put them at a disadvantage because it often proved problematic.

In those sections requiring diversion of the Des Plaines River, this work was the first priority of construction to clear the way for work on the Main Channel. In these sections and all others, the levee between the Main Channel and the Des Plaines River was a priority to protect the Main Channel work area from flooding by river overflow. The levee was given a completion date of December 1893, but few contractors met this date. Work progressed well on Sections 8 through 14 because of the uniform nature of the rock and because it was possible to work all year in dry conditions. In Sections 1 through 7, the glacial drift overburden would freeze in winter, and excavation was so difficult that work would cease in these months. In some sections, there were pockets of highly organic soils. These were difficult to excavate as dry material and were unsuitable for filling in the levee.

Removal of the overburden was accomplished with the use of scrapers and wagons for the first several feet of depth. Horses or mules supplied the motive power, and the wagons would make the round trip from the excavation area to the spoil area. Beyond a certain depth, this method became inefficient because of the slope of the haul road out of the pit. At this point, inclines were used where the soil was loaded on hopper cars, and the cars were pulled out of the pit by steam-powered locomotives or hoists. The hopper car was dumped and returned to the pit. Steam shovels were the equipment most often used to load the hopper cars. With a sufficient depth of working face of the material to be excavated, the stream shovel was the most efficient device. Most of the equipment was track mounted, and locomotives or teams of horses or mules were used to move the hopper cars.

Once the surface of the bedrock was exposed, the channeling machines were used to cut the vertical face of the channel walls. These operated on compressed air or steam and were mounted on tracks that paralleled the wall face. The cutting rod oscillated in the vertical plane for a depth of up to 16 feet. After each wall face was cut for a sufficient distance, compressed-air drills were used to make holes for dynamite charges in the rock mass to be excavated. After blasting, manual labor or steam shovels were used to clear away the rubble, now using hoppers for lifting and moving the rock to the spoil area.

One method used on many sections was the Lidgerwood travelling cableway, which consisted of a tower on each side of the channel with cables suspended between them supporting a trolley. Each tower was mounted on rails that ran parallel to the channel. The trolley itself was a hoist to raise and lower a pulley and hook. The trolley was pulled by cables back and forth across the channel, picking up hoppers full of rock and moving the hoppers to the spoil pile, where the hopper was dumped. Only one hopper could be moved at a time. The cableway was also used to move equipment and materials to or from the channel bottom. A power plant adjacent to one tower supplied the motive force for pulling the cables. The cableway allowed for the disposal of spoil on either side of the channel.

Another method used the Hulett-McMyler conveyor or the Brown cantilever crane. These machines used a double-cantilever inclined truss mounted on rails on the top of the channel wall. The double-cantilever inclined truss was mounted on a center base and tilted, with the lower end of the truss cantilevered over the channel and the higher end cantilevered over the spoil pile. Cables ran along the underside of the truss to move a track-mounted trolley along the truss. Again, the trolley acted as a hoist to raise and lower a pulley and hook. The uses of the double-cantilever inclined truss were similar to the cableway. A power plant next to the center base supplied the motive force for moving the cables and propelling the base laterally along the channel. This method was limited to spoil disposal on only the side of the channel where the inclined truss was positioned.

In Section 14, the contractor used single- and double-arm rotating derricks mounted on rails on the top of the rock wall. For the double-arm derrick, each arm was positioned on opposite sides of the derrick base and power plant such that when one arm was over the channel, the opposite was over the spoil pile. Each arm had two positions where pulleys were mounted for vertical hoisting. Each arm could be raised or lowered to facilitate radial positioning of the hoists. This method was limited to spoil disposal on only the side of the channel where the derrick was positioned. Other uses of the derricks were similar to the uses of the cableway and inclined truss.

In one other section, the contractor used a method prevalent in the Earth and Rock Section and the Earth Section. A section of the rock wall was cut to allow a sloping surface from top to bottom, parallel to the wall. Two rail tracks ran up the slope, to the spoil pile, and up the spoil pile. In the bottom of the channel excavation, a loop of track was laid along the face where rock was being removed, connecting to the slope up the wall. Steam locomotives or mule teams were used to move hopper cars along the track on relatively level surfaces, and steam hoists were used to raise or lower cars on the steeper slopes up the wall and spoil pile.

In some sections where the native rock was of high quality and blasting with dynamite could be controlled to produce large rock pieces, the rock was quarried to meet dimension requirements for masonry stone. This rock was saved for building the masonry-retaining walls on top of the bedrock where the solid rock did not extend over the full depth of the channel. This masonry stone was also used for bridge abutments and piers and for repairing the rock wall where the native rock would spall or fail.

1893 Construction Season

Since the contracts were awarded in July 1892, most contractors used the balance of the year to mobilize. The first significant work was accomplished in 1893. In addition, the SDC had to mobilize a force of field engineers to oversee the work. Formal reports on the progress of the work did not begin to appear until July 1893, under the direction of Chief Engineer Randolph. In Sections 1 through 4, the disagreement over conglomerate had already stopped work. The contractors in Sections 5 through 9 were all behind due to insufficient equipment on-site and problems in removal of overburden. Sections 10 through 13 were proceeding satisfactorily, but the contractor on Section 14 had difficulty financing his equipment and getting started. Work in the early part of 1893 was also affected by lawlessness along the channel route. Persons seeking work but finding none were vandalizing contractor equipment. This vandalism caused the contractors to hire private guards who were ruthless in protecting their employers' equipment and the safety of working laborers. The violence became pronounced, and the state militia was called to maintain order. After the SDC retained its own police force in July 1893, the level of lawlessness was reduced and work proceeded without significant interruption.

1894 Construction Season

U. W. Weston was appointed superintendent of construction early in this season, and he began to file meticulous monthly reports of the work. Levee work was reported to be completed in March, and the work area was deemed to be secure from flooding of the river. Work on Sections 8 through 13 exceeded expectations, and forecasts were that four of these sections would be completed in 1895. The contractor on Section 14 was falling behind due to experimentation with various methods of excavation, which were not successful. Section 15 was awarded, and work began on the downstream end of the Rock Section. Section 1 was idle for the first half of the year due to forfeiture, award to a new contractor, and a lawsuit. Sections 3 and 5 through 9 had turnovers in contractors due to forfeiture and readvertisement or through assignment. A landslide in Section 5 caused a two-month delay. The other sections were showing improvement. Dredges were brought in on Sections 6 and 7 due to deposits of peat or wet organic material on top of the bedrock. The year ended with nine of the fifteen sections showing better than 50 percent completion.

1895 Construction Season

Section 10 was completed in September, 13 in October, 9 and 12 in November, and 11 in December. Section 14 was the most improved, finishing the year at 88 percent complete. A record was set by the Section 14 contractor with 86,400 cubic yards of rock excavation in April, the highest production in any month. Although started two years behind the others, Section 15 made a respectable showing in its first year, finishing 1895 with 47 percent completion. The bedrock was assessed to be poor for masonry, and the contractor was given an extra for the use of concrete in building retaining walls. Section 8 was far ahead, but due to a delay in obtaining rights to cross the Santa Fe Railroad property, the rock under the tracks could not be removed. Peat and organic soils that were used in the levee in Sections 6 and 7 ignited and burned. In addition to the delay caused to put out the fires, extras were authorized for the levee repairs. Main Channel excavation was largely completed, and retaining-wall masonry was well underway in Section 2 through 7. Section 1 lagged behind with only 48 percent completion by year's end.

1896 Construction Season

This was the year in which the work in all fourteen sections was to have been completed. By year's end all were completed except for Section 1 at 83 percent. Due to the inability to agree on a price for cement masonry walls in Section 14, a separate contract was let for these walls in May. Construction of the Lockport Controlling Works in Section 15 began in June. A major disaster occurred in Section 7 when a length of retaining wall failed as discussed earlier. After inspection by SDC engineers, the slide was removed and the wall rebuilt by the contractor under an extra. Now that most of the contracts were completed and the contractors were demobilizing, the SDC took over responsibility for drainage. A tunnel was driven through the rock plug under the Santa Fe Railroad tracks in Section 8, and a pumping plant contract was let for Section 14. The plant went into operation in August, pumping drainage water out of the Main Channel into the Des Plaines River.

1897 Construction Season

This was a year of reduced activity for the Rock Section. U. W. Weston resigned due to ill health, and T. T. Johnston became responsible for reporting monthly progress. Most of the channel excavation and retaining wall contracts were complete. Work on Section 1 was finally completed in October, including completion of a trestle for a temporary bridge over the channel for Willow Springs Road and the paved transition from the trapezoidal to rectangular channel prisms. In Section 8, work began on the two bridges over the Main Channel and Des Plaines River for the Santa Fe Railroad tracks. Retaining-wall construction was completed in Section 14 in July. Capacity was added to the pumping plant in this section with the relocation of pumps from Section 3. In June, work began on a new contract for the construction of foundations for the Bear Trap Dam and other details to finish the Lockport Controlling Works, including the retaining wall across the south end of the Main Channel. This work was 25 percent complete at year's end. In Section 7, dimension stone left in the channel bottom was cut to size and prepared for shipment and use in the masonry foundations of bridges for Summit-Lyons, Willow Springs, Lemont, and Romeoville Roads.

1898 Construction Season

Construction was increased in Section 8 with additional contracts being awarded. One was for the Lemont Road (Stephens Street) Bridge over the Main Channel, and the other included realignment of the Santa Fe Railroad tracks crossing the Main Channel and removal of the rock in the channel prism below the railroad tracks. Removal of the rock did not begin until the following year, but the Santa Fe Railroad Bridge over the Des Plaines River was completed in 1898. In Section 1, the Willow Springs Road Bridge was about 35 percent completed at year's end. Work on the Bear Trap Dam, retaining walls, and other items in Section 15 were nearly completed as well. However, it became apparent that the pumping plant in Section 14 was inadequate to handle all drainage, and the work at the Bear Trap Dam was sometimes threatened by high water. Another pumping plant on the sill in front of the sluice gates in Section 15 was put under contract early in 1898 and went into operation in June. High water was no longer a problem. Shipment of the cut masonry stone from Section 7 occurred in November and December via the I&M Canal or by railroad. Also, in midyear, the monthly reports were submitted by Chief Engineer Randolph. T. T. Johnston resigned and went into private consulting work.

1899 Construction Season

This was to be the last year for construction before the Main Channel became operational; however, no one knew that at the start. Except for the rock plug in Section 8, there was no impediment to the flow throughout the 14 miles of the Rock Section. Only three temporary trestles were in the channel prism waiting for the permanent road bridges to be completed. However, in midyear there was a noticeable quickening in the desire to wrap up construction by year's end.

In Sections 1 and 12, the Willow Springs Road and Romeoville Road Bridges were completed and opened in April. The temporary trestles were sold and removed by the new owners. Late in the year, the former contractor pump station foundations and sumps in Sections 3 and 4 were removed and filled with concrete. Pockets of clay were found in the foundation of the retaining walls in Section 6 and in the rock walls in Sections 12 and 13. These were excavated and filled with concrete late in the year. In Section 15, work on the Bear Trap Dam was completed in June, although mechanical details in the operation of the gate were worked on through the end of the year. The added pumping plant at the sluice gates was removed in May to be used elsewhere. The pumping plant in Section 14 remained in service until December 12, whereupon it was disassembled, the foundation removed, and the void filled with concrete by January 1900.

Section 8 became the focus of concern. The Santa Fe Railroad Bridge was completed in June and the Lemont Road Bridge in July. However, the Lemont Road approaches were held up by work on the railroad roadbed realignment, so the bridge was not opened to traffic. The original contractor on the rock excavation in the Main Channel requested additional compensation for the work due to the delay caused by waiting for permission to occupy the site. The dispute was resolved in the contractor's favor and work started in July, but blasting was not allowed due to potential damage to other structures nearby. This slowed progress. By November, the contractor had almost reached the bottom grade when water ponded in the bottom became an impediment. A temporary cofferdam was authorized, but it failed under increasing water levels in December. With the bottom only 0.4 foot above grade, the chief engineer declared substantial completion, and the contractor was ordered to remove all equipment from the channel. As for the Lemont Road temporary trestle, in the first week of January 1900, ice built up on the upstream face, and it was washed away. Grading of the approaches to the permanent bridge was completed by the end of the month.

The Bear Trap Dam had no opportunity for a test under operating conditions until the middle of January 1900. As water gradually rose on its upstream leaf, the dam and neighboring sluice gates held back the water in the Main Channel. The opportunity for an operating test came on January 17, 1900, at 11:05 a.m. when the gate was lowered to begin the operation of the Main Channel. This event is more fully described in chapter 12.

References

Hill, Charles Shattuck. *The Chicago Main Drainage Channel: A Description of the Machinery Used and Methods of Work Adopted in Excavating the 28-Mile Drainage Canal from Chicago to Lockport, Ill.* New York: The Engineering News Publishing Co., 1896.

Berton, Pierre. *Niagara: A History of the Falls.* New York: Penguin Books, 1998.

History and Heritage of Civil Engineering. ASCE, 1998.

SDC. Proceedings of the Board of Trustees of the SDC. 1892–1900.

BUILDING THE CANAL TO SAVE CHICAGO

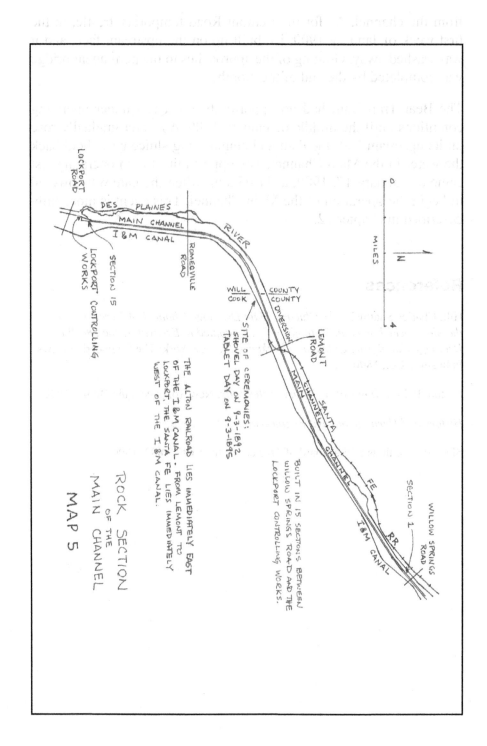

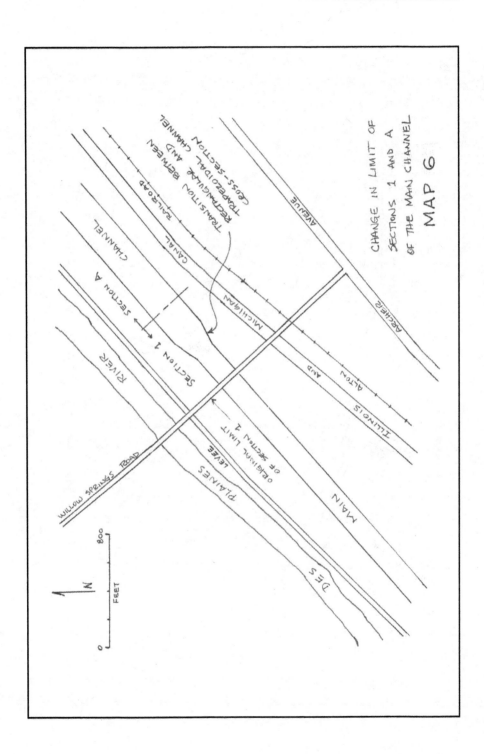

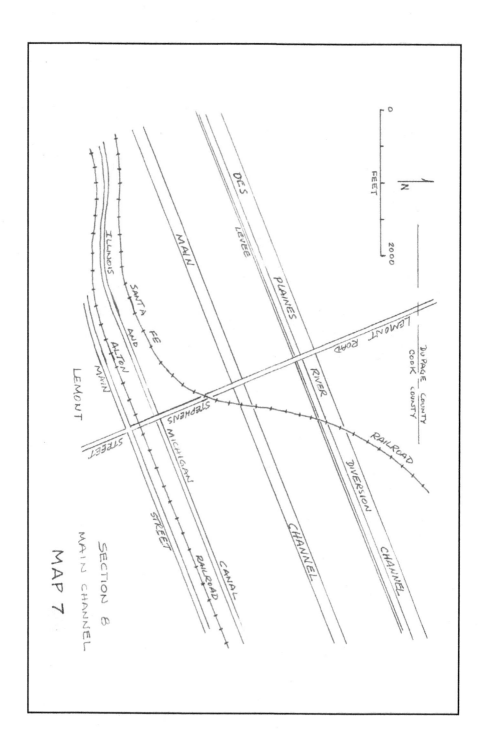

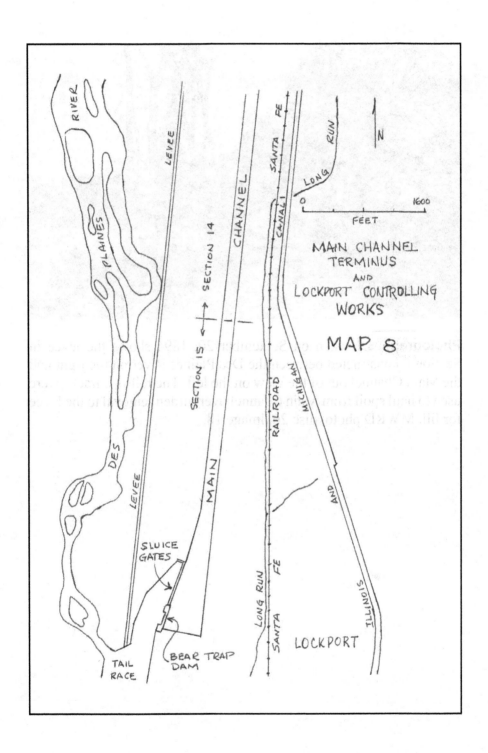

Photograph 5.1 taken on September 20, 1894 shows the levee in Section 6 constructed between the Des Plaines River on the right and the Main Channel out of the view on the left. The railroad tracks were used to haul spoil from Main Channel overburden removal to the levee for fill. MWRD photo, disc 26, image 18.

Photograph 5.2 taken in 1895 shows a hydraulic dredge being used in Section 6 to remove soft and wet soil overburden. The spoil slurry was used in the construction of the levee between the Des Plaines River and the Main Channel or it may have been placed where rock spoil would later be placed northwest of the Main Channel. MWRD photo, disc 5, image 14.

Photograph 5.3 taken on June 7, 1894 shows how manual labor was used to excavate overburden in Section 2. Spoil loaded into the hopper cars was then transported by horse or mule teams to the spoil pile on the northwest side of the Main Channel. Steam locomotives were also used to move the hopper cars. MWRD photo, disc 26, image 42.

CHAPTER 5: ROCK SECTION

Photograph 5.4 taken in 1895 shows a stream shovel in Section 5 used for overburden and rock excavation in constructing the Main Channel. In most section contracts, the steam shovel was the equipment used for excavation. These Bucyrus machines were the smaller forerunners of excavating machines used in the construction of the Panama Canal. The front of the steam locomotive used to transport the hopper cars to the spoil pile is shown at left. MWRD photo, Geiger set, image 219.

Photograph 5.5 shows a steam shovel at work in Section 15 in 1895 excavating the Main Channel. Solid rock was not near the surface at this location. Masonry or concrete walls were built to contain the flow in the Main Channel and excavated spoil was used as backfill behind the walls. MWRD photo, Geiger set, image 174.

CHAPTER 5: ROCK SECTION

Photograph 5.6 shows the bed rock surface sometime in 1895 in Section 5 after removal of overburden. Rock excavation would proceed until the full depth of the Main Channel was reached. Masonry walls were built on top of the rock to contain the eventual flow in the channel. MWRD photo, Geiger set, image 310.

Photograph 5.7 taken on June 10, 1896 shows a steam powered channeling machine used to cut the face of the vertical rock wall before drilling and blasting the rock mass for excavation of the Main Channel. This machine cut a neat vertical trench a few inches wide and up to 16 feet deep. The trench prevents cracking of the vertical rock face of the wall when the rock is broken up using dynamite for removal. The rock wall face is to the left and the rock to be blasted and removed is to the right. MWRD photo, disc 4126, image 81.

Photograph 5.8 taken on September 25, 1894 showing rock excavation of the Main Channel in Section 11. After the channeling machine cuts are made on each wall, holes are drilled in the rock and dynamite is inserted in each hole as shown here on the upper level. After the blast, the loose rock is removed from the lower level and placed in hoppers. The hoppers are lifted by overhead hoists and transported to the spoil piles shown in the left background. The rock is excavated in layers or lifts as shown here up to a total depth of nearly 30 feet. MWRD photo, disc 7, image 47.

Photograph 5.9 taken in 1896 shows workers loading broken rock by hand into a hopper in Section 5. The loaded hopper will be lifted by a hoist under a cableway and moved to the spoil area where it will be dumped. MWRD photo, disc 4126, image 87.

CHAPTER 5: ROCK SECTION

Photograph 5.10 taken on March 27, 1895 shows double-leg rotating derricks used to remove rock from the channel excavation in Section 14 on the left and deposit the rock on the spoil piles on both sides of the channel. Spoil was deposited on both sides of the Main Channel in this area. As can be seen, each leg of the derrick supported two hoists for hoppers. As two were being hoisted from the bottom of the excavation, two were being dumped on the spoil pile. MWRD photo, disc 3, image 95.

Photograph 5.11 shows the use of a double cantilever inclined truss in 1895 in Section 12 for rock removal from the excavation of the Main Channel. The Truss support was mounted on rails for travel alongside the Main Channel. Hoppers were hoisted from the floor of the Main Channel, moved by a trolley up the slope of the inclined truss and dumped on the spoil pile. In the foreground are gaps in the rock wall where the structural rock quality was poor or weak. The extra load of the cantilever incline may have contributed to the cause of the failure of the wall. MWRD photo, Geiger set, image 338.

CHAPTER 5: ROCK SECTION

Photograph 5.12 taken September 11, 1894 shows the removal of rock in layers or lifts beneath a double cantilever inclined truss in Section 12. A channeling machine is seen along the far wall under the farthest truss. Once the channeling machine completes the wall cut on both sides of the Main Channel, the rock will be drilled and blasted for rock removal. MWRD photo, disc 7, image 15.

Photograph 5.13 taken on August 31, 1895 shows the removal of rock under a cableway in Section 8. The contractor in each section could select their own machinery and execution of construction within the limits of the specifications. The cableway method employs two towers, one on each side of the Main Channel and outside the extent of spoil piles. Hoppers for the rock are suspended from a trolley traveling on the cableway between towers. This view also shows the clean vertical cut of the channeling machine in the far channel wall. The depth of each cut of the channeling machine was 16 feet. On the right workers are drilling holes for dynamite and on the left workers are loading blasted rock into hoppers. MWRD photo, disc 7, image 3.

CHAPTER 5: ROCK SECTION

Photograph 5.14 taken in 1895 shows the method used on Section 9 for removal of excavated rock from the Main Channel to the spoil area using hopper cars on tracks up a two-track incline on the side of the wall. The cars were pulled up the incline by cables and a steam powered hoist. After loading the hopper cars at the face of the blasted rock in the foreground, the loaded cars were pulled to the right and along the wall to the base of the incline. Empty cars were returned from the spoil area to the lower level on the other set of tracks on the incline and moved to the face of the blasted rock along the tracks shown to the left. MWRD photo, Geiger set, image 276.

Photograph 5.15 captures the instant of a blast of 500 pounds of dynamite on May 22, 1895 to fracture the rock for excavation of the Main Channel in Section10. Notice the rocks that are blown up into the air, requiring that all workers leave the area prior to the dynamite detonation. The contractor for this section also used the double cantilever inclined truss for lifting loaded hopper cars and transporting them to the spoil area. MWRD photo, disc 26, image 7.

CHAPTER 5: ROCK SECTION

Photograph 5.16 shows the nearly completed excavation of the Main Channel in Section 11 looking to the northeast from the top of the spoil pile near Romeoville Road. Although undated, the photo was probably taken in 1894 or 1895 based on the stage of completion. The double cantilever inclined truss was used for rock removal from the Main Channel in this section. Along the east wall two gaps are shown where the native rock was of poor quality. A masonry wall would be constructed to close this gap. The I&M Canal is shown to the right. The location of the Main Channel followed the route of the I&M Canal and was to the right of the I&M Canal in the downstream direction of travel. MWRD photo, disc 14, image 25.

Photograph 5.17 shows on March 27, 1895, in Section 13, three of the many gaps in the rock wall that had to be closed with a masonry wall. A steam powered derrick has been installed on top of the Main Channel wall to facilitate the construction of the masonry wall. MWRD photo, disc 3, image 92.

Photograph 5.18 shows the collapse of a cableway tower in Section 8 in 1895. The use of the cableway was shown in Photograph 5.13. In this view, one tower has collapsed toward the spoil pile and may have been caused by a break in the cable over the excavated area of the Main Channel. Also shown are parts of the steam boiler, mechanical equipment and the service building. MWRD photo, Geiger set, image 201.

Photograph 5.19 taken September 24, 1896 looking southwesterly in Section 5 shows the failure of the masonry wall and spoil pile. The cause of this failure was not determined, but it may have been influenced by an earthquake that occurred in the prior year. MWRD photo, disc 10, image 47.

CHAPTER 5: ROCK SECTION

Photograph 5.20 taken on September 22, 1896 shows failure of the rock wall beneath the masonry wall in Section 5. Perhaps the weight of the masonry wall above caused the rock to fracture and spall. The wall was repaired using masonry stone and mortar. MWRD photo, disc 4126, image 54.

Photograph 5.21 taken on May 17, 1895 shows the end wall of Section 7 adjoining Section 8. These walls separating each contract section provided for a defined separation and containment of drainage in each section. Each contractor was responsible for keeping the excavation as dry as possible. A cableway tower used by the Section 8 contractor is seen in the left background. Removal of the separation wall was the responsibility of the contractor who was the latest in completing work on the section. MWRD photo, disc 2, image 29.

CHAPTER 5: ROCK SECTION

Photograph 5.22 looking southwesterly shows the temporary trestle crossing for the Santa Fe Railroad in Section 8 in 1895. The rock wall left for the trestle also served as the temporary crossing for Lemont Road or Stephen Street. The original railroad embankment was replaced with this trestle so work could proceed with construction of the permanent bridge spanning the Main Channel and so rock excavation could proceed as close to the railroad crossing as possible. The original Lemont Road was about 500 feet southwest of the railroad crossing. MWRD photo, Geiger set, image 306.

Photograph 5.23 taken August 2, 1899 shows the Lemont Road or Stephen Street Bridge in the foreground and the Santa Fe Railroad Bridge in the background. The temporary trestle for the Santa Fe Railroad is shown atop the rock wall across the Main Channel. At this stage of construction progress the excavation work is complete and drainage of the excavation has been taken over by the SDC. The hole through the rock wall on the left allows accumulated water to pass downstream. The SDC operated a pumping station in Section 14. By the end of the year the new railroad bridge would be complete and in operation, and the temporary trestle and the rock wall removed. This view was taken from a temporary trestle installed across the Main Channel for Lemont Road. The new railroad embankment foreclosed the opportunity to route the temporary crossing for the road next to the railroad as shown in Photograph 5.22. MWRD photo, disc 9, image 28.

CHAPTER 5: ROCK SECTION

Photograph 5.24 taken on September 6, 1899 looking northeasterly shows the removal of the temporary trestle and the rock below the temporary crossing of the Santa Fe Railroad in Section 8 near Lemont. This view is from the top of the wall near the Lemont Road Bridge looking northeasterly. MWRD photo, disc 9, image 71.

Photograph 5.25 taken in 1895 in Section 5 shows the construction of the masonry wall on top of bedrock. A steam powered derrick in the Main Channel is used to lift and place the rock. Another derrick on top of the slope in the left background is also used for wall construction. A large group of spectators has gathered to watch the masonry wall construction process. MWRD photo, Geiger set, image 300.

CHAPTER 5: ROCK SECTION

Photograph 5.26 taken in September 1894 shows the construction of a rubble masonry wall on top of bedrock in Section 4. A single leg derrick is used to lift and place each masonry stone. The masonry courses were set back at vertical intervals to emulate the steps created by the channeling machines in solid bedrock. MWRD photo, disc 4126, image 88.

Photograph 5.27 taken July 7, 1896 in Section 4 shows another gantry-type piece of equipment for masonry wall construction. This equipment allowed for workers to set masonry stones at multiple levels as it travelled along the face of the wall. The horse in the foreground probably belongs to an SDC engineer or inspector, or possibly a supervisor for the Section 4 contractor. MWRD photo, disc 2, image 65.

CHAPTER 5: ROCK SECTION

Photograph 5.28 taken September 4, 1896 in Section 14 shows the construction of concrete wall. Cast-in-place concrete wall was used in this area due to the poor quality of bedrock and lack of a sufficient quantity of masonry stone. MWRD photo, disc 2, image 90.

Photograph 5.29 taken September 3, 1895, Tablet Day, in Section 10 at the Cook-Will County Line shows the celebration of the completion of excavation of rock in this and nearby sections. A large group of spectators has gathered to witness the placing of the granite tablet near the top of the northwest wall. MWRD photo, Geiger set, image 327.

CHAPTER 5: ROCK SECTION

Photograph 5.30 taken September 6, 1899 in Section 10 shows the granite tablet as it appeared before filling of the Main Channel with water. When filled with water, the tablet is visible above the normal water level. MWRD photo, disc 9, image 62.

Photograph 5.31 taken on May 5, 1899 from the Willow Springs Road Bridge looking northerly shows a crowd of spectators looking at the transition of the Main Channel from the vertical wall rectangular Rock Section to the sloping wall trapezoidal Earth and Rock Section. MWRD photo, disc 12, image 12.

CHAPTER 5: ROCK SECTION

Photograph 5.32 taken on July 20, 1897 from the Willow Springs Road Bridge looking northeasterly shows the transition of the Main Channel from the vertical wall rectangular Rock Section for contract Section 1 in the foreground to the sloping wall trapezoidal Earth and Rock Section for contract Section A in the background. This transition was necessary because bedrock was no longer prevalent toward Chicago. MWRD photo, disc 127, image 96.

Photograph 5.33 taken in July 1896 shows the SDC pumping plant in Section 14. Although the contractor for each section was responsible to maintain drainage of the excavation, this became burdensome for the contractors who finished early. The SDC assumed responsibility for drainage for the entire length of the Main Channel. The steam powered pumping plant was operated nearly to the end of 1899 when it was removed and the wall restored in time for filling of the channel in early 1900. MWRD photo, disc 2, image 66.

CHAPTER 5: ROCK SECTION

Photograph 5.34 taken on November 12, 1896 looking north from the terminal wall of Section 15 shows the gradual enlargement of the Main Channel for a turning basin in the foreground and construction of the Lockport Controlling Works sluice gates on the left. The bedrock was lower and walls were built to contain the water in the Main Channel. The water, if not contained, would flow over the landscape. MWRD photo, disc 10, image 61.

Photograph 5.35 taken on November 12, 1896 shows the fabricated sluice gates before installation in the guides on the inside face of each pier on the right. This view shows the downstream face of the gates. The upstream face is a flat plate. The area where the gates are standing would become the beginning of the tailrace channel leading to the Des Plaines River. MWRD photo, disc 10, image 62.

CHAPTER 5: ROCK SECTION

Photograph 5.36 taken on September 4, 1896 shows the construction of the Lockport Controlling Works sluice gate piers with the vertical gate guides. The Main Channel is to the right. MWRD photo, disc 2, image 89.

Photograph 5.37 taken on July 7, 1897 shows work nearly complete on the Lockport Controlling Works sluice gates to the right and excavation of the foundation for the Bear Trap Dam to the left. The view is from the Main Channel terminal wall looking northwesterly. MWRD photo, disc 4126, image 64.

Photograph 5.38 taken on September 15, 1898 shows the completed sluice gate structure to the right and construction of the foundation of the Bear Trap Dam in the foreground. The Main Channel is to the right. MWRD photo, disc 5, image 68.

Photograph 5.39 taken on February 7, 1899 shows the installation of the upstream (right) and downstream (left) leaves of the Bear Trap Dam. The downstream leaf is mounted on a fixed hinge along the downstream edge. The two leaves are hinged at their junction. Water will flow over the top of the leaves and by raising the leaves, the rate of flow will be diminished. MWRD photo, disc 7, image 101.

Photograph 5.40 taken February 7, 1899 shows details of the downstream leaf of the Bear Trap Dam. The holes in the foundation allow flow to enter or exit the chamber providing buoyancy to reduce the dead weight of the leaves and to facilitate raising or lowering the dam. MWRD photo, disc 8, image 3.

Photograph 5.41 taken April 28, 1899 shows the completed leaves and the installation of the hoists at each end of the Bear Trap Dam. MWRD photo, disc 5, image 11.

CHAPTER 5: ROCK SECTION

Photograph 5.42 taken May 5, 1899 shows an inspection group on top of the 160-foot long Bear Trap Dam. MWRD photo, disc 12, image 8.

Photograph 5.43 taken on June 30, 1899 shows the completed Bear Trap Dam in the center and the seven sluice gates in the right background. Without water in the Main Channel, the sluice gates and dam could not be tested. MWRD photo, disc 9, image 11.

CHAPTER 5: ROCK SECTION

Photograph 5.44 taken April 29, 1899 shows the left half of a two-photograph panorama of the Lockport Controlling Works. This view shows the Bear Trap Dam in the center and the first three sluice gates on the right. On the left and foreground are the walls of the end of Section 15 of the Main Channel. MWRD photo, disc 12, image 5.

Photograph 5.45 taken on the same day as Photograph 5.44 shows the right half of a two-photograph panorama of the Lockport Controlling Works. This view shows six of the seven sluice gates and the eight bulk-headed sluice gate bays. MWRD photo, disc 12, image 4.

Chapter 6

Earth and Rock Section

The Earth and Rock Section of the Main Channel extends from Willow Springs to Summit. Most of the materials encountered for channel excavation consist of glacial drift with some rock within the depth of excavation at a few locations in the reach. The channel route follows the route of the Des Plaines River and the I&M Canal in a northeast-southwest direction throughout the reach. At Summit, the route of the Des Plaines River turns north, separating from the routes of the Main Channel and I&M Canal, which turn more easterly toward Chicago's downtown (see map 9).

Introduction

After the award of the initial fourteen contracts for the Rock Section, the board took action to proceed with work for the Main Channel upstream of Willow Springs. In August 1892, the board decided to split the work upstream of Willow Springs into two segments, namely Willow Springs to Summit and Summit to Chicago. This decision was based on two factors. First, the work just put under contract downstream of Willow Springs required control of floods on the Des Plaines River to prevent excess flows that could inundate the construction work. Therefore, work along the river upstream of Willow Springs needed to be undertaken as expeditiously as possible. Second, the board still had not decided as to the route of the channel upstream of Summit. Orders were issued to proceed with land acquisition for the ROW between Willow Springs and Summit and to prepare plans and specifications with the expectation that contracts could be awarded by November 1892.

Route Selection and Design

Within a week of the above order to deliver plans and specifications, Chief Engineer Williams presented the board with a dilemma. Due to the lack of information from soil borings, it was thought that the Main Channel upstream of Willow Springs would only need to be designed for a capacity of 5,000 cubic feet per second (cfs) because rock was not known to be near the surface and this capacity would comply with the act. However, soil borings performed in the spring of 1892 revealed otherwise, and the chief engineer had instructed his staff to provide for a capacity of 10,000 cfs. This change took the board by surprise as some felt strongly that the lower capacity would suffice. There also was concern for the increased cost of a larger channel. The chief engineer provided alternate specifications for both capacities but recommended the large capacity. A motion to advertise for the large capacity failed to pass. A motion to advertise both alternates did pass.

While the alternates were out for bid, the board requested a legal opinion on the matter of Main Channel capacity. In his October 1892 Opinion on Rocky Stratum, Attorney Orrin Carter supported the large capacity for several reasons, including the rock near the surface, population, and navigation. Meanwhile, the bids were opened, and the board approved the award of Sections D, E, and F. Sections A, B, and C were rejected due to unacceptably high prices for excavation. However, the board could not decide which alternate capacity to designate, and the matter was referred back to committee. A majority of the committee, and eventually the board, favored the large capacity, and this was approved in November 1892.

The Main Channel route would lie northwest of the I&M Canal and southeast of the Des Plaines River, with diversion of some portions of the river a necessity. The reach was divided into six contracts, Sections A through F. The length of each section varied from 4,500 to 8,000 feet, with all six sections totaling 32,500 feet or 6.2 miles. Section A began just northeast of the Willow Springs Road crossing, abutting Section 1, and ended near the present-day LaGrange Road. Sections B, C, and D spanned the distance to the Chicago Terminal Railroad crossing. Section E began at and included the crossing of this railroad and ended at the old Summit-Lyons Road crossing. Section F began at and included the crossing of the old Summit-Lyons Road and ran to the Range Line, or what is now Harlem Avenue.

The three contracts for Sections D, E, and F were awarded to the same contractor. However, the contractor failed to qualify for the contract for Section E, and the bid was rejected and readvertised with the contracts for Sections A, B, and C. These bids were acceptable, and awards were made in February 1893.

To provide for the Main Channel, the Des Plaines River was relocated in a diversion channel throughout most of the Earth and Rock Section. The diversion channel begins as soon as the Des Plaines River flows under the Santa Fe Railroad tracks, just south of today's Forty-Seventh Street and west of Harlem Avenue. To separate the flood waters of the river in the diversion channel from the Main Channel construction, a levee was also constructed. This levee permanently separates the Des Plaines River from the Main Channel even today, over 116 years later. Today the levee runs from this location to Romeoville Road and is designated the Centennial Trail.

Contract Requirements

Provisions of the contract documents were similar in many respects to those for the Rock Section. There were some significant differences in the specifications due to the different work conditions. Item 4, Cross Section for Walled Channel, went into detail regarding the method of excavating the rectangular prism, with the only difference being a vertical limit of 12 feet for the offsets rather than 16 feet as in Sections 1 through 15. Despite these provisions, the Main Channel prism in Sections A through F was entirely trapezoidal in shape, and no walled channel was built.

Item 6, Cross Section for Unwalled Channel, required that where the bottom of the channel was wholly in glacial drift, the bottom shall be

210 feet wide with side slopes of 2 horizontal to 1 vertical. Where the bottom was wholly or partly in rock, the same width and slopes apply to the channel above the rock to give a uniform cross section, and the rock shall be excavated to the bottom grade with side slopes of 1 horizontal to 2 vertical. The rock shall be excavated without use of channeling machines, but the surface was to be free of sharp and ragged projections.

The transition between the rectangular and trapezoidal prisms was covered in Item 7, Connection of Walled and Unwalled Channel, which actually became part of Section 1. The transition called for a gradual enlargement to increase the bottom width from 160 to 210 feet over a distance of 125 feet. The retaining wall was to gradually decrease the slope of its channel face from the near vertical and change to slope paving as it approached the 2 horizontal to 1 vertical side slope of the unwalled channel. This transition remains to this day visible from the Willow Springs Road Bridge. In the empty channel, it presented a remarkable sight.

Item 8, River Diversion, required that the river diversion channel was to be constructed in Sections A, B, C, E, and F. The diversion channel size and slopes were the same as in the Rock Section. Item 10, Levees, indicated that a continuous levee was to be constructed between the Main Channel and the Des Plaines River from the Santa Fe Railroad upstream of Summit to Willow Springs. The levee was to have a crest width of not less than 20 feet, a crest height of 20 feet above datum, side slopes of 3 horizontal to 2 vertical, to be watertight, and to be completed by March 1, 1893.

Under Item 15, Highways, the contractor in Section E was required to construct an embankment from the I&M Canal to the northwest side of the Des Plaines valley paralleling the existing Summit-Lyons Road for the new road. The embankment was to have a crest width of 80 feet, a crest height 30 feet above datum, side slopes of 3 horizontal to 2 vertical and was to be graded at the two channels to accommodate bridge abutments.

The time for construction and completion was specified in Clause G, requiring completion of the work to be ready for inspection as provided in the statute by April 30, 1896.

Des Plaines River Spillway

While the diversion channel for the Des Plaines River and the levees were under construction, it was necessary to limit the flood flows in the river downstream of Summit. The problem caused by excess Des Plaines River flows toward Chicago was explained in chapter 1. To accomplish this, a levee and spillway were built near present-day Forest View so that excess flood flows could be diverted through the Ogden-Wentworth Ditch to the Chicago River. Fortunately, a major flood did not occur during the period of construction, or Chicago might have been devastated again, as it was in 1885. Upon completion of the diversion channel, the crest of the spillway was eventually raised to prevent flood water from heading toward Chicago at any time (see map 10).

Plans and specifications were presented to the board by Chief Engineer Randolph in October 1893 and referred to committee. It was Randolph's recommendation that the work on the spillway be accomplished using SDC forces under the supervision of a competent foreman. The spillway was 400 feet long and had a crest 16.25 feet above datum and was constructed on bedrock on the east bank of the Des Plaines River. Levees were also built along the east bank flanking the spillway. The crest height would prevent any flow toward Chicago when the Des Plaines River flow is less than 4,167 cfs. Without the spillway, flows could begin to flow toward Chicago when the river reached a discharge of 833 cfs. This spillway and the improved river diversion channel downstream would afford a much greater degree of protection for Chicago.

The board approved the recommendation at the next meeting, and Randolph came back to the board with an agreement for the work in December, which was approved. The chief reported completion of the spillway in October 1894. Remnants of the spillway abutments are evident today along the levee, about midway between Joliet Road and Forty-Seventh Street.

Summit-Lyons Conduit and Levee

In the purchase of ROW, a requirement was included in the sales contract for certain properties to the northwest of the river diversion channel in Sections D, E, and F, which required the SDC to construct what is now the Summit-Lyons Conduit and levee. These lands were low and poorly drained, and construction of the diversion channel would not improve the drainage. The levee had to be built when the river diversion work was underway, but the drain or conduit did not have to be built until the Main Channel was completed. This drain improved the usefulness of the low property because at this location, the Des Plaines River was much higher than the Main Channel, and the river would not provide a good outlet for the drainage. The construction of the levee was ordered by the board in May 1893 to be built by the contractors for Sections D, E, and F.

Diversion Channel and Levee Height

In September 1893, the change of grade for the Main Channel was approved as explained in chapter 5. This change affected the Earth and Rock Section as well, and change orders were issued to the contractors. The levee separating the river and diversion channel from the Main Channel tied into the embankment of the Santa Fe Railroad upstream of Section F. The diversion channel began as soon as the river passed under the Santa Fe tracks near present-day Riverside, cutting off a wide bend in the river that reached nearly to the I&M Canal. The levee top was 20 feet above datum, placing it about 15 feet above the top of the bank of the Main Channel.

Changes in Sections A and B

Within the first year, some changes were coming to Sections A and B. In October 1893, to expedite the completion of the levee in Section A, the board authorized an extra for the construction of a 4,500-foot-long timber pile trestle along the centerline of the levee and additional compensation for fill brought in by rail from Section 1. This accommodation was made because of a lack of suitable material for the levee within Section A. The Section A contractor also requested additional compensation for excavation below the grade line of the diversion channel, which was claimed to be unavoidable due to the use of a dredge. Use of a dredge was the most expedient means of excavation due to the peat and wet organic soils. Meanwhile, the Section B contractor requested extra compensation for the excavation of unanticipated hard material found in the river diversion channel. In November, Chief Engineer Randolph recommended methods to deal with these soils found in excavating the Main Channel in Sections A and B.

All the issues dealing with these two sections were referred by the board to a subcommittee consisting of the attorney, the chief engineer, and the superintendent of construction for a single comprehensive recommendation. Meanwhile, the contractor for Section 1 was delaying the prosecution of work and had not responded to an order to resume work. The contract was proceeding to forfeiture and to readvertisement. The subcommittee reported in February 1894, recommending several changes based on savings in the cost of construction. First, the peat and wet organic soil in the upper-channel prism of Sections A and B, found to be from 8 to 20 feet thick, would be excavated by hydraulic dredge beyond the limits of the side slopes, and hard soils from the lower-channel prism would be deposited as backfill to meet the limits of the side slopes. These measures would provide for a stable channel prism. If a sufficient quantity of material was not available in the section, it would be made available from the new contract for Section 1.

Second, the Section A contract calls for the construction of 2,700 feet of retaining walls, but recent borings indicated that much less wall would be needed. Since suitable rock material was not available in Section A but was available in Section 1, the extent of wall needed and a transition from the rectangular to trapezoidal channel prism, some 700 feet in length, should be added to Section 1 and deducted from Section A. Third, since sufficient quantities of suitable material for the levee are not available within Section A, an appropriate length of levee should be added to the new Section 1, where suitable material is available. Fourth, the Section A contractor should be granted additional compensation for excavation below the grade line of the diversion channel provided all other claims are withdrawn. This additional excavation was judged to be a benefit, and the contractor had shown diligence in the prosecution of the work. The board approved these recommendations and the amendment of the contract for Section A.

Difficult Materials in the Glacial Drift Controversy

This subheading was the name given to one of the major problems affecting work in this reach of the Main Channel. A similar problem as that with conglomerate in the glacial drift in several contracts in the Rock Section occurred also in Sections C through F. It began with reports of construction in February 1894 that some contractors had encountered hard materials that could not be excavated by scrapers or steam shovels without the use of explosives. Petitions were submitted in March and April requesting reclassification of this material as solid rock for an extra for additional compensation. Three petitions were submitted to the board and one to Chief Engineer Randolph.

Board members had visited the work sites in February as a result of the changes in the contracts for Sections A and B and had requested Randolph to respond to the petitions, which he did, rejecting the requests. The contractors again laid their demands before the board, and in May, the board requested Randolph to form a commission of his staff to investigate and report back on the nature of the difficult material. Test pits were opened by the contractors in Sections E and F to examine the material. Information for Section D was available in construction reports, and existence of the materials in Section C was suspect. Commission members submitted individual reports on the nature of the materials, with some members also suggesting cost adjustments. Based on the test pits, it was estimated that there was 1,000,000 cubic yards of hard material and 400,000 cubic yards of quicksand. The board took the individual commission member reports under consideration and reported in July with majority and minority recommendations.

The majority recommended that the petitions be denied based on their belief that the difficult nature of the material was not demonstrated and that it was included in the glacial drift definition of the contract. The minority recommended that a special committee consisting of the president and two trustees he selected be appointed to arrive at terms of settlement and report at the next meeting. The minority was persuaded that the material on Sections E and F was indeed difficult and that to summarily dismiss the petitions may jeopardize progress in construction of the Main Channel. The board adopted the minority recommendation.

The Special Committee on Settlement of the Difficult Material Controversy reported back in August with majority and minority recommendations. The majority recommended that because the committee members could not agree, the matter be referred to the chief engineer for resolution. The minority, being Trustee Cooley, recommended in a very lengthy and passionate report that the contractors for Sections E and F be awarded additional compensation based on the demonstrated difficulty of the material and that there be no consideration given for the other two sections as the evidence of difficulty was lacking. The board adopted the majority recommendation, and Randolph subsequently denied the petitions for extra cost.

In view of the earlier decision in the changes for Sections A and B, this decision appears abrupt by comparison. The consequences of the decision reached in August 1894 were that the contractors for Sections C and D agreed to proceed with the work and, later, were given extensions in time in exchange for a waiver of claims. The contractors for Sections E and F refused to proceed and were declared in forfeiture. The work was readvertised and awarded to other contractors at slightly lower unit costs for glacial drift. Extras were eventually awarded for rock revetment or slope paving in Sections E and F, where the quicksand was encountered.

Bridges

Only two routes traversed the Earth and Rock Section: the Summit-Lyons Road and the Chicago Terminal Railroad. Bridges were needed for these two routes over the Main Channel and Des Plaines River diversion channel. Temporary trestles were constructed by the contractors so that excavation of the Main Channel and diversion channel could proceed. Permanent bridges were built under separate contracts. The permanent bridges for Summit-Lyons Road over the river diversion channel and the Main Channel were completed in August 1898 and June 1899, respectively. For the Chicago Terminal Railroad, the bridges over the Main Channel and the river diversion channel were completed in October 1898 and November 1899, respectively.

Contractor Continuity and Completion of the Work

Although the work was to have been completed by April 1896, none of the contractors finished on time. Sections A, B, and D were finished in 1897 by the second contractor for each section. Section C was finished in 1898 by the original contractor. In 1899, Section E was finished by the fourth contractor and Section F was finished by the third contractor.

Construction Progress and Methods

The award of contracts late in 1892 and early 1893 resulted in a start of work in the spring of 1893. Work progressed with reports back to the board at monthly intervals. Chief Engineer Randolph reported in November 1893 that work on the levee was essentially complete, although the levee in some sections would be the source of problems. Individual contractors were given time extensions for the resolution of their problems with the levee. Early in 1894, the ends of Sections A, D, E, and F were changed. The reason for the change in the boundary between Section A and Section 1 was explained earlier in this chapter. For Sections D, E, and F, the change was for the purpose of making the end coincidental with a bridge crossing the Main Channel. By July 1894, it was reported that progress in most sections was behind. The progress of work in this reach was plagued by peat and wet organic soils, levee failures, disputes with contractors, and failure of some contractors to make reasonable progress.

Completion of the diversion channel was the first priority, and this work appeared to proceed without problems, probably because the work could be accomplished in the dry. Except for Section D, diversion of the river was necessary throughout the Earth and Rock Section. Initially, reports of construction indicated successful completion of the levee. These reports were based on initial observations of the levee reaching its full size and height. However, with soft materials in Sections A and B, problems appeared in sloughing and settlement, which caused failure of the levee. Since the initial excavation of the Main Channel occurred in the former riverbed over much of the length of this reach, the contractors had to deal with peat and wet organic materials. This gave rise to the use of dredges in three of the sections.

Once the excavation could be accomplished in the dry and with the prevalence of glacial drift, construction proceeded similarly to the work in Sections 1 through 7. Scrapers and wagons pulled by horses or mules were used for the first several feet of excavation. Next, steam shovels were used to load hopper cars, with the cars pulled to the spoil area by locomotives. The locomotives were also used to move equipment around as most were mounted on steel wheels and rails. With greater depth of excavation, inclines with steam hoists were used to move the excavated material out of the cut and up to the top of the spoil pile. Some contractors used more elaborate conveyors, trusses spanning the spoil pile, or cantilevered sloped inclines. These methods and equipment were used as well for the small amount of rock excavation. Placement of the rock revetment paving was accomplished with steam-powered derricks.

1893 Construction Season

Although work began early in the year, by the summer, Sections A, C, and D were behind schedule. In particular, work on the levees in Sections A and C was so far behind that a recommendation was made to hire additional workers. This was not acted upon. Hydraulic dredges began removal of peat and wet organic soils in Sections A and B. By the end of the year, the Summit Spillway was complete, and levees in all sections were sufficiently underway as to alleviate concerns for flooding of the Main Channel excavation areas.

1894 Construction Season

A cold winter idled the dredges until March. In addition to Sections A and B, dredges began excavation in Sections C and D. Unstable soils, peat, and organic soils caused sloughing and settlement in the levees on Sections A and B, and pervious materials in the levees in Sections C and E caused excessive leakage. Sections A, C, and E were flooded. To assist in the construction of the levee in Section A, the contractor was given an extra to construct a causeway on timber trestles along the centerline of the levee. Railroad tracks were mounted on the causeway, from which fill material was dumped to form the levee.

The contractors on Sections B and D isolated their work areas with additional levees so that dry excavation could proceed. Solid bedrock was encountered in Section D, where it had not been anticipated in preconstruction surveys. In Sections C through F, hard materials were found in the drift, prompting the use of explosives and claims for additional compensation. This dispute resulted in work stoppage in Sections E and F. Work resumed in Section F, but Section E was forfeited, readvertised, and awarded to another contractor in September. By year's end, completion ranged from a high of 56 percent in Section F to a low of 25 percent in Section E. The other sections were between 41 and 47 percent.

1895 Construction Season

Extreme cold in January and February virtually shut down all work except for Section D. Peat and wet organic soils in Sections A and B caused slow progress, resulting in the use of human and animal power rather than heavy machinery. These sections were sufficiently dewatered to allow some areas to be excavated in the dry by midsummer. Section E experienced equipment problems in the wreck of a truss conveyor, the burning of a dredge, and the mechanical failure of an incline conveyor. Section F was awarded to a second contractor early in the year. An interior levee was removed contrary to the advice of the resident engineer, and a flood on the river inundated the work in December. At year's end, Sections A and E were only 50 percent complete, Section F was 66 percent, and Sections B, C, and D were 74 percent complete.

1896 Construction Season

Another cold winter resulted in shutdown of the work in February. Despite this shutdown, the contractor on Section A made remarkable progress compared to the others. Rock was encountered on Section B in the bottom of the excavation where none was expected. All glacial drift materials were removed by September in Sections B and D. In Section D, a cableway was installed to handle rock removal. Section E again experienced equipment problems with the collapse of an incline in September and destruction of an equipment shed in October. The contractors in Sections E and F encountered financial difficulties late in the year, with the Section E contractor going into receivership and the Section F contractor abandoning the work. The SDC had to pay the contractors' employees in both cases. These sections stood at 69 and 81 percent completion, respectively, at year's end, whereas Sections A, B, C, and D stood at 87, 99, 93, and 91 percent completion, respectively.

1897 Construction Season

Sections A through D saw completion of work by fall. Despite new contractors in both Sections E and F, only modest progress came to channel excavation, finishing the year at 73 and 88 percent completion, respectively. Under extra work, the contractors in Sections D and E installed temporary trestles for the railroad and road crossings, levee repairs were accomplished in Section E, and revetment paving was begun in Sections E and F. Equipment problems plagued Section F in this year, with the collapse of a truss conveyor and a steam shovel being demolished when it tipped over and rolled downslope to the bottom of the excavation.

1898 Construction Season

Work continued in both Sections E and F with glacial drift removal from the Main Channel prism at the railroad and road crossings, rock removal, and slope paving. Extra work for the construction of the permanent bridges over the Main Channel and the river diversion channel and embankments for the railroad and roadway were also completed. The contractors for the permanent bridge substructures and superstructures began their work. The railroad bridge over the Main Channel was completed, and the temporary trestle was removed. The Main Channel excavation work stood at 91 and 99 percent completion for Sections E and F, respectively.

1899 Construction Season

Main Channel excavation was completed in January on Section F and in August on Section E, making this 6.2 miles of Main Channel ready for the reversal of the Chicago River. The Summit-Lyons Road permanent bridge was also completed, and the temporary trestle was removed. The eastern (upstream) end of Section F in the Earth and Rock Section abuts Section G in the Earth Section. As explained in the next chapter, the Earth Section was constructed to a lesser capacity. The contractor on Section F completed the slope paving to the end of Section F, where a transition to Section G occurs.

References

SDC. Engineering Department Annual Report for 1899. Proceedings of the Board of Trustees of the SDC. 1899.

SDC. Proceedings of the Board of Trustees of the SDC. 1892–1900.

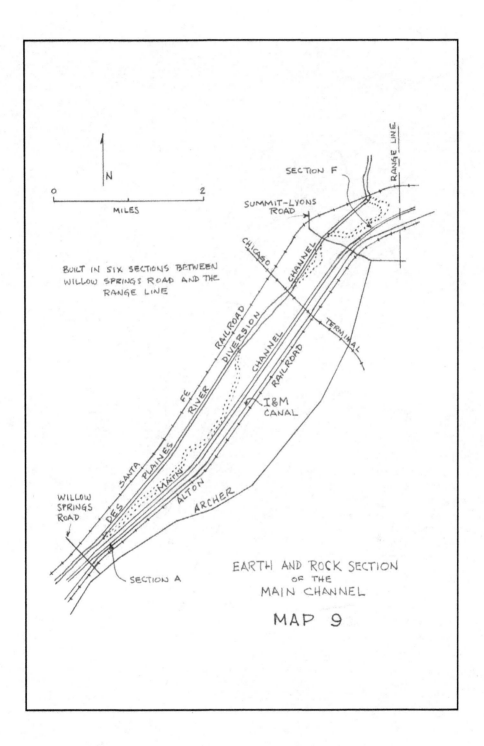

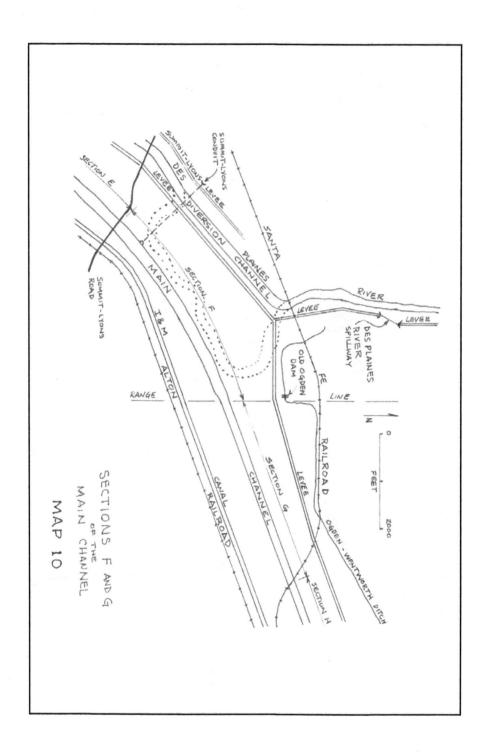

Photograph 6.1 taken on May 31, 1894 shows a hydraulic dredge in Section A used to excavate peat and organic wet soils in the overburden. The dredge spoil slurry was pumped to diked areas between the construction site and Des Plaines River to form part of the levee. MWRD photo, disc 3, image 13.

Photograph 6.2 taken on May 31, 1894 shows a dipper dredge used on Section A to remove deeper soft and wet overburden. The dredge spoil was cast to the side or loaded into scows and used in the levee between the construction site and Des Plaines River. MWRD photo, disc 3, image 19.

CHAPTER 6: EARTH AND ROCK SECTION

Photograph 6.3 taken on July 10, 1894 shows the use of mule-drawn scrapers to remove unconsolidated dry soils in Section D. The excavated material was loaded into mule-drawn carts for transport to the spoil area or to the levee constructed to separate the Des Plaines River and the Main Channel. MWRD photo, disc 26, image 30.

Photograph 6.4 taken on May 28, 1894 shows the removal of overburden in excavating the Main Channel and use of the spoil in constructing the levee to the left between the Main Channel and the Des Plaines River in Section F. MWRD photo, disc 3, image 8.

Photograph 6.5 taken on June 30, 1896 shows the excavation of the Main Channel in Section A. The steam shovel in the lower right is digging into the face of the soil and loading the spoil onto hopper cars that are then towed to and up the incline for transport to the spoil pile. MWRD photo, disc 2, image 56.

Photograph 6.6 taken on June 30, 1896 shows the same incline as in the prior photograph and the truss spanning the spoil area in Section A. The hopper cars are towed up the incline and across the truss to the point of dumping the spoil. MWRD photo, disc 2, image 59.

Photograph 6.7 taken on September 18, 1896 is a closer view of the steam power plant, incline and truss in Section A shown in the prior two photographs. Here a hopper car is shown shortly after depositing its load of spoil under the truss. MWRD photo, disc 1, image 16.

Photograph 6.8 taken in 1895 shows the use of a steam shovel for excavating the Main Channel in Section D. MWRD photo, Geiger set, image 275.

CHAPTER 6: EARTH AND ROCK SECTION

Photograph 6.9 taken on April 20, 1895 shows large boulders discovered in the overburden during excavation of the Main Channel in Section B. The contractor claimed an extra for removal of these boulders resulting in a dispute between the SDC and the contractor. MWRD photo, disc 2, image 13.

Photograph 6.10 taken on October 12, 1894 shows the pumping plant used by the contractor in Section B to keep the excavation de-watered. MWRD photo, disc 7, image 68.

Photograph 6.11 taken on October 6, 1898 shows final grading and slope paving or armoring for the Main Channel in Section F. This protection was necessary to resist the erosive force of the water that would eventually flow in the Main Channel and to maintain the stability of the sloping channel bank. MWRD photo, disc 5, image 90.

Photograph 6.12 taken on April 13, 1899 shows the finished channel with slope paving in Section F. The paving was accomplished with limestone slabs from rock excavation. MWRD photo, disc 11, image 82.

CHAPTER 6: EARTH AND ROCK SECTION

Photograph 6.13 taken on April 13, 1899 shows the transition in channel size at the division between Sections F and G, the current location of Harlem Avenue. This view is from the Summit-Lyons Road Bridge looking northeasterly and shows the removal of the temporary bridge to the right. Notice that the slope paving on the right bank extends farther upstream to the limit of Section F while the transition occurs within the end of Section F on the left bank. The smaller channel was widened early in the next Century to meet the requirement of the enabling act as the population grew. MWRD photo, disc 4126, image 100.

Photograph 6.14 looking north shows construction of the Des Plaines River Spillway in early 1894. The river is on the left behind the trees at the base of the moraine. This location is about 0.25 miles north of present day 47th Street. Concrete or masonry was used to form the spillway and wing walls. Earthen levees flanked the spillway on each side to separate the Des Plaines River from the Chicago Lake Plain, the broad flat landscape that stretched to Chicago and Lake Michigan. The smaller earthed dikes on each side of the spillway were temporary and served to keep the construction site dry. Photo from Brown, G. P., *Drainage Channel and Waterway*, Chicago, R.R. Donnelley and Sons Co., 1894.

Photograph 6.15 taken on March 6, 1894 shows the completed spillway looking north allowing a small flow over the spillway toward Chicago. Small flows were beneficial in improving water quality and reducing nuisance odors in the West Fork and South Branch while the Des Plaines River diversion channel and Main Channel were being constructed. The levees flanking the spillway between the river and Chicago Lake Plain served to control excess flows from causing overland flooding that would eventually be channeled in the Ogden-Wentworth Ditch, West Fork, South Branch and Chicago River. Photo from "Lecture" found in the Greeley and Hansen, Chicago, archives. "Lecture" was prepared by the SDC in June 1918.

Chapter 7

Earth Section

The Earth Section of the Main Channel extends from the Range Line at Summit, now Harlem Avenue, to the West Fork of the South Branch just west of Robey Street, now Damen Avenue, in Chicago. There is no bedrock in this section, and most of the excavation was in fairly uniform deposits from the Chicago lake plain. The deposits resulted from the various stages of Lake Chicago, which existed here as the glacier was retreating to the northeast. This reach was also notable because of the many railroad crossings (see map 11).

Decisions on Route Selection

The route of what would be the Earth Section was actually one of the early considerations of the board when, in June 1890, Chief Engineer Lyman E. Cooley was directed to undertake work to present the board with not less than four routes of a channel between the South Branch and Summit. The intent was to build a channel over these 8 miles and to build two pumping stations, one to lift the waters of the South Branch into the new channel, and the other to lift the water out of the other end of the new channel into the Des Plaines River. The board's consideration of and decisions on these early routes, resulting in the selection of a route and ROW ordinance, is explained in chapter 4. The routes are shown on map 4.

Due to concerns over the cost of land along Route 4 and objections raised by the railroads, the board requested a review as it appeared that use of the I&M Canal would be much less costly. Chief Engineer Williams reported to the board in June 1892, following a review of the routes between the South Branch and Summit, strongly favoring the I&M Canal Route 3 over the earlier selection of Route 4. Without adopting this new route, the board rescinded the ROW ordinance adopted earlier for Route 4. Both President Wenter and Trustee Cooley advocated for the I&M Canal route in October 1892. After award of the first fourteen contracts for the Rock Section, Wenter expressed his opinion that use of the I&M Canal would be more expedient, allowing for completion of the entire channel by 1896.

Chief Engineer Williams confirmed this in another report the same month and also addressed the issue of the Stock Yards and South Fork. The initially favored Route 4 would have originated on the South Fork at Thirty-Ninth Street to remove the filth from the Stock Yards. With favor now directed toward the I&M Canal, Williams recommended an open channel from the lake along Thirty-Ninth Street to the South Fork so that half of the dilution water from the lake could flow through the Chicago River and South Branch and the other half through the new channel and South Fork. Cooley, while supporting use of the I&M Canal for reasons of cost and time, was not impressed with Williams's plan. He favored instead the construction of lakefront interceptors and filling in the South Fork, replacing its drainage function with a closed conduit. None of these other ideas was pursued, however, but use of the I&M Canal was approved, and a ROW ordinance was adopted in November 1892.

The decision came none too soon for some. A spectrum of organizations led by the Knights of Labor had been discussing the matter. They convened a conference at the Grand Pacific Hotel in early November, from which came objections to Route 4, suggesting use of the I&M Canal instead but questioning the need for such a large channel. The critics also sought the means for immediate relief for the Chicago River. The summer of 1892 had been hot and dry, exacerbating the conditions of the river. It was more the stench of the river, not the threat to the water supply, that gave rise to public concern. The board answered, happily, that the route of the I&M Canal did offer the least costly route.

The enlarged I&M Canal and the larger channel downstream of Summit was needed because the existing I&M Canal simply did not have sufficient capacity. It was a popular misconception that if the lock gates at Lock No. 1 at Lockport were opened, allowing unrestricted flow, the Chicago River would be reversed. This same misconception gave rise to criticism in 1871 when the city of Chicago constructed the Bridgeport Pumping Plant at the head of the I&M Canal. As to relief for the Chicago River, the board was of the opinion that the use of the I&M Canal and the channel construction downstream of Summit was the most expedient path to relief.

Bids for dredging the I&M Canal were opened in January 1893 and confirmed the estimates for enlargement. The use of the I&M Canal with a large pumping station at Summit was submitted as an official proposal to the I&M Canal Commissioners. The board was advised that the I&M Canal property was available to the SDC at no cost, and the route offered less problems with railroad crossings since these railroads already crossed the I&M Canal.

In March, the I&M Canal Commissioners notified the board of their intent to submit to the SDC their own proposal for use of the I&M Canal, that the governor was not committed to such use of the canal by the SDC, and that the interests of the state must be protected. The board and the commissioners met on the matter without resolution, and later, the commissioners, without submitting their proposal to the SDC, requested that they receive for their review the SDC's plans and specifications for the dredging. The attorney for the commissioners submitted an affirmative opinion on the use of the I&M Canal by the SDC. The board authorized transmittal of the dredging plans and also submitted a rebuttal to the opinion.

Reaching a point of frustration, the board adopted and transmitted a resolution demanding a response from the I&M Canal Commissioners by the end of April 1893. This demand was met with rejection by the commissioners and a further legal opinion that required SDC payment of tolls for use of the I&M Canal, allowed the SDC to charge fees for use of docks built by the SDC, placed the SDC interests subordinate to the I&M Canal Commissioners, and required the SDC to be responsible for the maintenance of navigational capability.

The board requested another review by their attorney. In response, the SDC attorney rendered an opinion that a limited capacity channel and pumping station was contrary to the statute, suggested that the Illinois attorney general be approached regarding the interests of the state, and recommended a lawsuit as the only way to resolve the impasse with the commissioners. Further, the attorney was uncertain as to the outcome of a lawsuit and advised that it was certain to delay the construction. In May 1893, with desire to proceed unfettered by the I&M Canal Commissioners, the board reconsidered and selected a route that would provide for a new large-capacity channel connecting to the West Fork at Robey Street, a route that had not been considered previously. The board took these actions: use of the I&M Canal was rescinded, preparation of plans and advertisement for bids was ordered, and a ROW ordinance was adopted.

Channel Design and Contract Award

The reach was divided into eight sections for construction. The reach from the Range Line at Summit to Corwith, a former railroad junction, now Central Park Avenue, was divided into Sections G, H, I, K, L, and M. Oddly, there was no Section J, and the reason for its omission was not given. The length of these sections varied from 2,853 to 5,480 feet, for a total of 25,782 feet or 4.9 miles. It was intended that these sections would be excavated similarly to those west of the Range Line, that is, in the dry with the spoil being deposited parallel to the channel on extra ROW. Plans and specifications for Sections G through M were advertised in September 1893, and the six contracts were awarded in December 1893.

Having had the experience of awarding twenty contracts, the board expressed its opinion "that it is not always wise to let to the lowest bidder" and that "the Board must exercise a wise discretion and select such bidders as in its opinion will prosecute the work diligently to completion within the limit of time set forth in the contract." They also concluded "that there is a substantial advantage in working the sections in groups of two" adjacent sections for one contractor. It just so happened that the lowest bidders fell in this pattern, except for Section G. However, the low bidder on Section H lowered his bid on Section G to be the same as the lowest bidder for this section, and the award was made on that basis.

Section G begins at the Range Line (now Harlem Avenue) and ends at the Santa Fe Railroad crossing and includes the crossing in the section. Section H extends from the Santa Fe Railroad crossing to the extension of present-day Menard Avenue (5800 West). Section I extends from Section H to the extension of present-day Laramie Avenue (5200 West). Sections H and I do not include any rail or road crossings. Section K extends from Section I to the east line of the ROW of the Belt Line Railroad crossing and includes the crossing. Section L begins at the Belt Line Railroad crossing and ends at Crawford (present-day Pulaski) Avenue. Section M extends from Crawford Avenue to the west line of the ROW of the Santa Fe Railroad Twenty-Sixth Street Line (later the Illinois Northern Railway), or the extension of present-day Central Park Avenue. Sections L and M do not include any rail or road crossings.

The reach from Crawford Avenue to Robey Street was divided into Sections N and O, being 4,506 and 7,078 feet in length, respectively, a total of 11,534 feet or 2.2 miles. These sections were to be excavated by dredges with the work proceeding from east to west. In addition, these two sections included two road and three railroad crossings. The advertisement for Sections N and O occurred in February 1894, and awards followed in May. Section N begins on the west line of the Santa Fe Railroad Twenty-Sixth Street Line crossing and ends at the east line of the Chicago, Madison, and Northern Railroad (later the Illinois Central Railroad) crossing. In addition to those two railroad crossings, the section includes the Kedzie Avenue crossing. Section O ends at the centerline of Robey Street on the West Fork and includes the Western Avenue and Southwest Boulevard road crossings and one crossing in which three different railroads were adjacent. Plans and specifications for Sections N and O were advertised in February 1894, and the two contracts were awarded in May 1894 (see map 12).

The contracts for Sections N and O were advertised with alternatives providing for disposal of the excavated soil in spoil piles along the ROW or providing for removal of the spoil and disposal as the contractor saw fit. Upon award of the contracts, the board indicated that removal of the spoil is preferred due to the eventual enlargement of the channel to meet the requirement of the statute as the population grew. More importantly, the spoil was needed as fill for local streets and railroads. To facilitate construction of the Main Channel, numerous streets and railroads required rerouting and raising the grade of the approaches to bridges over the channel.

In addition to the need for the Collateral Channel, as explained below, it was necessary to remove a rock ledge in the bottom of the West Fork near Campbell Avenue (2500 West) for the passage of dredges and barges. Although not specified in the contract, the low bidders indicated in their bids that excavation by dredging was less expensive than digging in the dry. The contractors offered to remove the rock ledge so that the dredging could proceed at three working faces—namely, west from Robey Street and both east and west from the Collateral Channel—rather than from one face at Robey Street. This offer was attractive to the SDC for reasons of less cost and time. The two contractors with the lowest bids agreed to work together with three dredging contractors who were also in the bidding. The agreements for Sections N and O were written, naming all five companies.

The plan for excavation of Sections N and O by dredging implied a scheme for eventual connection of the West Fork to the Main Channel and the flow reversal of the Chicago River and South Branch. An earthen plug or dike would be left in place at the line separating Sections M and N. Sections N and O would be dredged and, with standing water, would be an extension of the West Fork. Sections M through 15, excavated in the dry, would remain empty until filling via some introduction of water through the plug at the line between Section M and N. This plan did not materialize due to complications brought on by eventual construction of the Eight-Track Railroad Bridge for the three railroads.

Collateral Channel

Included in Section O was the construction of the Collateral Channel at Albany Avenue, connecting the Main Channel with the West Fork. This adjunct, some 1,600 feet in length, was necessary to obtain agreement from three railroads for a new bridge to accommodate the railroads' crossing of the Main Channel. The railroads already operated and maintained a movable bridge over the West Fork, which was about 2,000 feet north of the proposed new bridge over the Main Channel. Movable bridges so close together on these three busy railroad lines would result in intolerable delays. The railroads would not agree to the new bridge. Riparian owners of docks along the West Fork, west of the railroad line, demanded that navigation be maintained. To obtain the railroads' agreement and to satisfy the riparian owners, the SDC agreed to build the Collateral Channel. The existing railroad bridge over the West Fork could remain and become unmovable when the new movable bridge over the Main Channel went into service. Further, navigation to the West Fork, west of the railroad lines, would be available through the Collateral Channel.

The location of the Collateral Channel was not determined prior to the award of the contract for Section O in May 1894. However, the ROW was purchased in September, and its location was fixed in October. The contractor proposed a method of excavation at less cost than the bid price, and this was accepted by the SDC. The Collateral Channel would prove to be crucial in the filling and operation of the Main Channel in early 1900.

Contract Requirements

The contracts for Sections G through M were all identical and were similar in format to the earlier contracts; however, there were some differences due to the nature of the work. The alignment of the channel in Item 1, Location, was referenced to the I&M Canal, the centerline being 663 feet northwesterly of and parallel to the north reserve line. Item 4, Dimensions of Cross Sections, requires the channel to have a bottom width of 110 feet and side slopes of 2 horizontal to 1 vertical. This size is based on a capacity of 5,000 cfs, which is all that the statute required at this time based on population. There were no references in the specifications to walled channels, retaining walls, changes in the I&M Canal, or the Des Plaines River diversion channel.

Item 6, Levee, required the construction of a continuous levee throughout all six sections, parallel to the channel, along the northerly ROW boundary, with a top width of 15 feet and a top elevation that varied uniformly from 20 feet above datum at the west end to 16 feet above datum at the east end. No time limit was placed on the completion of the levee, but it was "to be undertaken immediately and completed as soon as practicable." The levee in this reach protected the channel from high water in the Ogden-Wentworth Ditch, the remnants of Mud Lake and the West Fork. Item 7, Disposition of Material, placed greater controls on the spoil piles built up along the channel for disposition of the excavated material. The toe of the spoil was to be no closer than 80 feet from the top of the channel bank.

There no longer being a difference in the type of soil excavated, Item 9, Classification of Material, simply referred to excavation. Glacial drift and solid rock was no longer defined. Item 10, Quantity and Quality of Material, placed greater responsibility on the contractor to take the risk for knowledge of the material to be excavated. Time, Clause G, required the work to be completed and ready for inspection by May 31, 1896.

The contracts for Sections N and O, as indicated above, were written in the names of five contractors. Many provisions are the same as in earlier contracts, but there are not only some significant differences but differences between N and O as well. The location of the channel in both Sections N and O, similar to Sections G through M, is referenced to the north reserve line of the I&M Canal. Item 6, Levee, in both agreements calls for a levee along the north ROW border from the west end of Section N to the Pan Handle Railroad embankment. Oddly, there are numerous references to this railroad name, but this name is different than the names of the three railroad companies the SDC signed agreements with for use of this ROW. Pan Handle was apparently a common name applied to the ROW occupied by the three railroad lines. The levee top elevation descended from 16 feet above CCD at the west end of Section N to 12 feet above CCD at the railroad embankment.

In the agreement for Section N, there are Items 7 and 7a, but they have no title. Item 7 requires that the westerly 100 feet of the channel not be excavated until so ordered by the chief engineer. Assuming that the excavation was going to be performed by dredge, this unexcavated material would act as a plug, holding back the flow of the South Branch until all work was complete. Item 7a required that all excavated material in the section not needed for leveling up the surface of the adjacent lands, for levees, or for street and railroad approaches was to be removed so that the ROW may be unencumbered and ready for any use to which it may be designated for the good of the SDC. It also referred to possible limitations in the work as may be imposed by the City of Chicago, Board of Park Commissioners, railroads, or as may relate to grade adjustment of streets or railroads. Clause G, Time, required the completion of work for inspection by May 31, 1896, and for the levee work to be completed by October 1, 1894. A new clause, N, Contractor's Bond, set forth the requirements of a bond. Although a new clause, a bond was always required in all previous contracts per language in the advertisement.

In the agreement for Section O, Item 7, Collateral Channel, required the excavation of a channel from the West Fork to the Main Channel at a location specified by the chief engineer, a bottom width of 60 feet at 12 feet below datum, and side slopes of 1 vertical on 1.5 horizontal. Item 7a, West Fork Navigation, required the removal of material to provide adequate depth for passage of dredges and barges at no additional cost to the SDC. It did not refer to this material as *rock*. The item continued with the same language in Item 7a of the agreement for Section N.

Excavating Machines

Construction of the Earth Section saw some innovations in machinery for moving the excavated material to the spoil piles paralleling the route. The uniformity of the soils in the Chicago Lake Plain allowed for the machines to be simpler than those for more complex materials. In most of the sections where excavation was in dry conditions, methods and machinery used were similar to those used on the Earth and Rock Section. Steam shovels were used more than any other type of equipment. To move the excavated material to the spoil pile, conveyors or track-mounted hopper cars were used. The material was lifted to the top of the spoil pile by inclined structures. Motive power was supplied by steam-driven hoists.

The contractor on Section H used a double-cantilever conveyor, a bridge-type structure that spanned the horizontal distance between spoil piles on each side of the channel bank. It was commonly called the Hoover and Mason conveyor after the two engineer-mechanics who designed it. The machine was mounted on rails on the top of either bank for movement along the channel. Material was excavated in the channel bottom and loaded onto the conveyor. The conveyor moved continuously to lift the excavated material from the channel bottom to the top of the spoil pile. A coal-fired steam engine supplied the motive power to move the conveyor and the machine along the track. The machine was moved along track as the excavation progressed.

The Section H contractor erected this monstrous apparatus. It was a truss bridge spanning the channel with cantilever arms projecting over the spoil area on each end. The horizontal distance between the ends of the arms was 640 feet. At the end of each cantilevered arm was a large-diameter drum over which a steel belt passed. The vertical distance from the bottom of the channel to the top of the arm was 90 feet. Both the bridge and the arms were parallel trusses, and the belt ran between the trusses, supported on timber cross members at each truss panel point. The belt consisted of a series of steel pans linked together as a chain, which travelled on a track mounted on the arms and bridge. The sag of the belt between ends went to the bottom of the excavation. Excavated material was loaded on to the pans by steam-driven plows. The steam-driven belt would lift the material to the end of the arm, where it was then dumped onto the spoil pile.

The conveyor began operation in September 1894 and ran successfully through a series of trials of different methods used to load the material onto the conveyor. The device was designed to be self-loading, but in some situations this feature did not work well, and scrapers or steam shovels were needed to loosen the material. On November 8, 1894, the conveyor suffered a disabling accident. The accident was caused by the breaking of one of the timber cross members on the north arm. Since these supported the track and conveyor belt, the belt sagged and the load on the north arm increased. Before the drive mechanism could be disengaged, the north arm buckled and collapsed. Once the wreckage was cleared away, a steam shovel was brought in to continue excavating the face using track-mounted hopper cars and locomotives to remove the spoil. Prior to the accident, the excavator had shown promise in its efficiency compared to other machines, reaching an output of 940 cubic yards per 10-hour shift. An electric-light plant was installed to allow operation for two shifts daily.

Repair of the conveyor was begun immediately and progressed through the winter. However, as the repairs were nearly completed in January 1895, disaster visited again. The entire conveyor with cantilever arms spanned the channel and spoil areas with the only point of contact with the ground being two sets of wheel trucks on tracks on the top of each side of the channel banks. On the morning of January 21, 1895, a strong wind prevailed along the channel, and due to inadequately secured chock blocks on the tracks at the trucks, the wind propelled the entire structure along its tracks for some distance until, reaching the end of the track, the leading trucks fell to the ground and the entire structure overturned, completely wrecking both cantilever arms. The main truss landed on its side and was little damaged.

The contractors vowed to repair the conveyor as soon as possible and return it to service. However, to continue the work in the meantime and to comply with concern for progress expressed by the chief engineer, they proceeded to install the tried-and-true steam shovels and inclines. The Hoover and Mason conveyor was returned to service in June 1895 and continued to perform without mishap until August 1896. However, it never performed as well as it did in the fall of 1894 and could not better the performance of the steam shovel and incline.

Bridges

The number and design complexity of the bridges was perhaps the most challenging aspect of the Earth Section. There were two street bridges, Kedzie Avenue and Southwest Boulevard/Western Avenue. These were built as conventional center-pier swing bridges. From west to east, the railroad bridges were the Santa Fe (6400 West), Belt Line (4600 West), Santa Fe (3600 West), Madison and Northern (3100 West), and the Eight-Track (2500 West). The Eight-Track Bridge accommodated three railroads: the Northern Pacific, the Pittsburgh and St. Louis, and the Stock Yards. All railroad bridges were designed and constructed as center-pier swing bridges, except for the Eight-Track. The three railroads would not agree to the swing design. In 1898 they did agree to the design of four double-track, single-leaf bascule bridges.

In addition to the above bridges over the Main Channel, the proximity of the Madison and Northern Railroad and Kedzie Avenue required the construction of a grade separation with the railroad on a viaduct over Kedzie Avenue. Completed and in service prior to the end of 1899 were the two Santa Fe Railroad Bridges, the Madison and Northern

Railroad Bridge and viaduct, and the Kedzie Avenue and Southwestern Boulevard/Western Avenue Bridges. The Belt Line and Eight-Track Bridges were completed after the Main Channel was placed in service in early 1900. All the railroad bridges remain in service as of 2011, although they are no longer movable. The two street bridges have been replaced. More detail on the design and construction of the bridges is found in chapter 10.

Construction Methods and Contractor Continuity

Aside from the ill-fated Hoover and Mason conveyor, work on the Earth Section proceeded smoothly, although not on schedule. The reason for most of the delay was due to achieving agreement from the railroads on ROW access and bridge design. In Sections G through M, excavation typically proceeded with the use of scrapers to remove the upper layers, followed by use of steam shovels and inclines to complete the channel prism. Although dredging was to be used in Sections N and O, some delays in acquiring ROW forced the use of dry methods early on in Section O. In Section N, dredging never was used because of inaccessibility to dredges through Section O. A large portion of the excavated material was deposited in local roadways to raise their grade and in low areas to promote land development. Western Avenue, Southwest Boulevard, and Kedzie Avenue are some of the streets whose grade was raised using spoil. Another large amount of spoil went to Douglas Park for landscaping. The spoil from dredging was removed by scow, towed up the South Branch and out the Chicago River to Lake Michigan, and deposited in open water or was used to fill in Lake Front Park, now Grant Park. Some dry spoil also went to Lake Front Park via the Illinois Central Railroad.

Unlike the contracts for work downstream of Summit, all the original contractors finished their work without forfeiture or serious disputes. Completion by the end of May 1896 was required, but none of the contractors met this date. However, the contractors for Sections I through M completed their work in September or October of that year, except for the crossings of the Belt Line and Santa Fe Railroads. Sections G and H were completed in August 1897, except for the Santa Fe Railroad crossing. The work on Sections N and O went right up to the end of 1899 due to delays caused by bridge construction and by the final excavation and letting water into the entire Main Channel in January 1900.

1894 Construction Season

As soon as weather allowed, work began on Section G through M with setting up construction camps and removing upper layers with scrapers to build the levee while inclines were being installed. Once the steam shovels and inclines were in place, work proceeded expeditiously, except for Section H where the contractor was preoccupied with the Hoover and Mason conveyor. Start-up on Sections N and O did not occur until midyear. Dry excavation proceeded on both, with the spoil going into the levee and street filling. On Section O, the dock wall and piling was removed west of Robey Street so that the dredges could begin their work. In October, the dredging subcontractor, who was supposed to dispose of spoil in the lake, was found by the US Army Corps of Engineers to be dumping spoil in the river. The contractor was ordered to remove the dumped spoil.

1895 Construction Season

A cold winter shut down work for dry excavation until March and dredging until May. Once begun, work proceeded apace, with Sections I through M finishing the year with over 90 percent completion. Section G was 71 percent complete at the end of the year. Section H was only at 56 percent completion due to the problems with the Hoover and Mason conveyor. Little work was accomplished on Section N, and the year ended with only 13 percent completion. In Section O, dredging was completed from Robey Street to Western Avenue by September but could not go much farther due to lack of ROW from the railroads. Dredging began on the Collateral Channel in June, and the section was 44 percent completed by year's end. The temporary trestle for Western Avenue was installed.

1896 Construction Season

Excavation of the Main Channel continued on Sections G and H and, except for the railroad crossings, neared completion by the end of year. Little work remained to finish up Sections I through M, except for the railroad crossings. The contractors worked on final grading, a few slope failures, and drainage. In Section O, the temporary trestles for the railroad crossing were installed, and dredging of the Collateral Channel was nearly completed. Dredging of the Main Channel proceeded west of the temporary railroad bridge. However, movement of scows under the temporary bridge and frequent delays due to changes in the temporary bridges caused dredging progress to be slow. Sections N and O finished the year at 32 and 66 percent completion, respectively.

1897 Construction Season

Sloughing of the slopes in Section G was pronounced with the spring thaw. It was resolved by excavating a trench on the top of the slope to a depth that penetrated the hard clay above a layer of soft material. This trench relieved the soft material and the pressure on the slope face. The slopes were restored. Excavation of the channel prism in Section G was completed, and work began on the substructure of the Santa Fe Railroad permanent bridge. In Section H, final grading of the channel bottom and surface drainage were completed. All machinery was removed from the completed Sections I and K. Significant progress was made in Section N with excavation of the channel, removal of spoil to Douglas Park via a temporary railroad line, raising of several railroad embankments, excavation of the Kedzie Avenue subway beneath the Madison and Northern Railroad, and the installation of temporary trestles over the Main Channel. Dredging of the channel prism in Section O neared completion, and an extra was granted for widening the basin west of Robey Street. Construction of the substructure of the permanent bridge for the Southwest Boulevard/Western Avenue was begun. Sections N and O finished the year at 60 and 84 percent completion, respectively. The slow pace of the contractor on Section N was an ongoing problem.

1898 Construction Season

Except for the two railroad crossings, excavation of the channel prism in Sections G through M was complete, and the contractors only had to maintain drainage via pumping. Incomplete work downstream of Summit prevented drainage by gravity down to Section 14. Work on Section N was off to a slow start due to cold, rainy weather and a brief labor strike. Excavation of the channel prism with the spoil removed by rail to Douglas Park continued throughout the year until completion in November. Work progressed on temporary trestles and permanent bridges for Kedzie Avenue and the two railroads. In Section O, dredging was completed to the west end by May but continued to September in widening the basin west of Robey Street. Additional temporary trestles were installed at the eight-track crossing to accommodate the swing bridge design, and work commenced on the permanent bridge for Southwest Boulevard/Western Avenue.

1899 Construction Season

The permanent bridge for the Santa Fe Railroad in Section G was completed in February, and the excavation of the channel prisms beneath the railroad crossings in Sections G and K were both completed in August, allowing drainage by gravity to Section 14. Drainage of Sections M and L by pumping to the West Fork was completed in August. Two temporary trestles were built for the Belt Line Railroad in Section K, being completed in May and December. The first one was rendered unusable by a change in the permanent bridge design and was removed in January 1900. The substructure of the Belt Line Railroad permanent bridge was completed in December. The Main Channel was now completed from Section M to Lockport.

Much was left to be completed in Sections N and O. In Section N, permanent bridges over the channel were completed by May for the two railroads and one street. Excavation of the channel resumed in April and was completed in December, including the prism at the three bridges. Dredging of the channel prism in Section O had been mostly completed, but a change was ordered to provide for connection of the West Fork to the Main Channel. Due to delays in the construction of the Eight-Track Railroad Bridge, the channel at the bridge location would have to be dewatered for substructure construction. In addition, the eventual flow of water through the channel beneath the bridge would not be acceptable to the bridge construction activity. It was decided to let water in to the Main Channel through the Collateral Channel, thereby allowing most flow to bypass the bridge site. In August, construction of an earth-fill dam across the channel east of the Eight-Track Railroad crossing was begun. It was completed in November, and the water west of the dam pumped out to the West Fork. Substructure work was commenced immediately.

The permanent bridge for Southwest Boulevard/Western Avenue was completed in May, but the Eight-Track Railroad bascule bridges were far from completion. The substructure for these bascule bridges was completed in late December. Also, in November and December, a wooden flume was built into the remaining earth at the south end of the Collateral Channel. The flume would be used to control the flow of water into the excavated channel. A mound of earth at the north end of the flume held back the water in the Collateral Channel, which was connected to the West Fork. The West Fork and Collateral Channel would be the route for the water to fill the new channel. The filling of the Main Channel is described in chapter 12.

References

Hill, Charles Shattuck. *The Chicago Main Drainage Channel: A Description of the Machinery Used and Methods of Work Adopted in Excavating the 28-Mile Drainage Canal from Chicago to Lockport, Ill.* New York: The Engineering News Publishing Co., 1896.

SDC. Proceedings of the Board of Trustees of the SDC. 1892–1900.

CHAPTER 7: EARTH SECTION

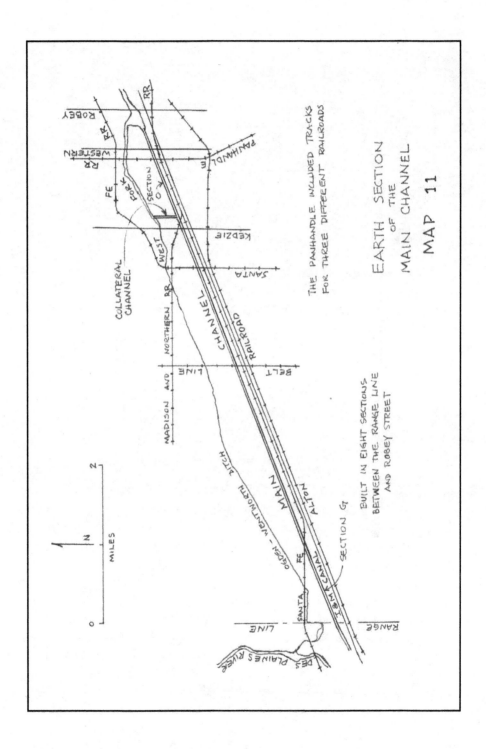

BUILDING THE CANAL TO SAVE CHICAGO

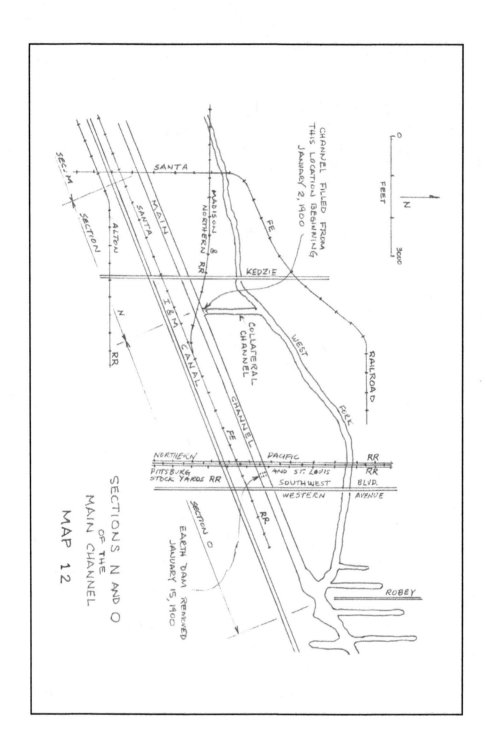

Photograph 7.1 taken in 1895 shows a steam shovel excavating the Main Channel in Section G. The excavated material is dumped into a hopper above a conveyor belt hidden behind the steam shovel. MWRD photo, Geiger set, image 204.

Photograph 7.2 taken in 1895 shows the conveyor belt, incline and steam power plant in Section G used to transport the excavated material from the steam shovel to the spoil area. MWRD photo, Geiger set, image 205.

CHAPTER 7: EARTH SECTION

Photograph 7.3 taken in 1895 shows the truss spanning the spoil area in Section G. The excavated material in the Earth Section was of uniform quality without boulders. Even though various mechanical devices were used to excavate and transport soil, the steam shovel was the most-used type of equipment. MWRD photo, Geiger set, image 203.

Photograph 7.4 taken on September 15, 1894 shows the Mason-Hoover excavating machine used in Section H. The narrower width of the Main Channel in the Earth Section allowed for this machine to span the area to be excavated. It rode on rails on each side of the Main Channel and the extended arms on each end cantilevered over the spoil areas. A continuous belt from end-to-end of the structure was used to transport the excavated soil to the spoil area at either end. This view is from the Santa Fe Railroad tracks looking northeasterly. MWRD photo, disc 7, image 27.

Photograph 7.5 taken in 1895 shows the Mason-Hoover excavating machine following an accident in Section G in which one of the extended arms collapsed. Despite the touted efficiency of this machine, it could not exceed the performance of the steam shovel. This view is from the Santa Fe Railroad tracks looking westerly. The water in the foreground is probably a channel constructed by the contractor for drainage. MWRD photo, Geiger set, image 181.

Photograph 7.6 taken on June 11, 1894 shows the excavation method used on Section I, which is similar to the method used on many other sections. This contractor used a double track incline. MWRD photo, disc 26, image 41.

CHAPTER 7: EARTH SECTION

Photograph 7.7 taken in June 1894 shows another view of the excavating machinery and method used on Section I. The hopper cars are lifted up the incline by a steam powered hoist. MWRD photo, disc 3, image 50.

Photograph 7.8 taken in 1895 shows the excavation of the Main Channel in Section L. The total depth of the excavation is accomplished in two lifts. The upper lift spoil is transported to the left and the lower lift spoil is transported to the right. MWRD photo, Geiger set, image 212.

Photograph 7.9 taken on October 14, 1896 shows a completed Main Channel excavation in Section L. Contractors who completed their portion of the Main Channel excavation were required to keep the excavation de-watered or nearly so until all work was completed for the entire Main Channel. This was eventually found to be a hardship for those contractors who finished early and the SDC assumed the responsibility for drainage. MWRD photo, disc 4126, image 49.

Photograph 7.10 taken on June 28, 1899 shows the channel excavation nearly completed and spoil piles in the Earth Section. MWRD photo, disc 8, image 100.

CHAPTER 7: EARTH SECTION

Photograph 7.11 taken in 1895 shows a steam shovel excavating the Main Channel in Section M. MWRD photo, Geiger set, image 208.

Photograph 7.12 taken in 1895 shows the two track incline for hopper cars used in excavating the Main Channel in Section M. The steam power plant used to move the hopper cars is at the base of the upper incline and part of a steam shovel is shown at work in the foreground. MWRD photo, Geiger set, image 207.

Photograph 7.13 taken in 1895 shows the upper incline used for transporting excavated material to the spoil pile in Section M. MWRD photo, Geiger set, image 209.

Photograph 7.14 taken on July 7, 1894 shows a dipper dredge used to excavate the Main Channel in Section O. Being an extension of the South Branch, this part of Section O was excavated by floatable equipment because it was more economical than building a coffer dam, dewatering and excavating in the dry. The portion of Section O east of the Eight-Track Railroad Bridge was the only part of the Main Channel completely excavated in this manner. The spoil was loaded into scows, towed to the Chicago Harbor and used to make land along the lakefront in what is now Grant Park. MWRD photo, disc 3, image 77.

CHAPTER 7: EARTH SECTION

Photograph 7.15 taken July 7, 1894 is another view of the dipper dredge and scow shown in Photograph 7.14. Dipper dredges were used to excavate the Main Channel in Section O. Due this part of Section O being an extension of the South Branch, excavation by floatable equipment was more economical than building a coffer dam, dewatering and excavating in the dry. The portion of Section O east of the Eight-Track Railroad Bridge was the only part of the Main Channel completely excavated in this manner. The spoil was loaded into scows, towed to the Chicago Harbor and used to make land along the lakefront in what is now Grant Park. MWRD photo, disc 26, image 35.

Photograph 7.16 taken October 17, 1894 shows another view of excavation in Section O. Due to the lack of more sophisticated equipment for under water excavation, the banks of the Main Channel were left without graded side slopes. In time, these banks would develop a trapezoidal cross-section shape. MWRD photo, disc 7, image 74.

Photograph 7.17 taken in 1897 shows a tug boat used to move scows in the dredging in Section O. Loaded scows were towed to the lake and the spoil was deposited for fill in what is now Grant Park. MWRD photo, disc 4126, image 53.

Photograph 7.18 taken on September 11, 1899 shows the nearly completed excavation for the Main Channel in the western portion of Section O. This portion was excavated using steam shovels and the spoil was transported from the site and used as fill for roads and park landscaping. MWRD photo, disc 9, image 74.

Chapter 8

Joliet Project

The reach from Lockport to Joliet was initially included in the consideration for the first set of contracts. However, its character was much different than the Main Channel route upstream of Lockport, and it was omitted by only awarding contracts for channel excavation to a point upstream of Lockport. Near Lockport, the Des Plaines River began a relatively steep descent to Joliet, at some places flowing over exposed dolomite. Once in Joliet, the river was joined by the I&M Canal, and there had been constructed dams to create navigation pools and locks to navigate between pools. Private interests had also constructed water power mills and electrical generation plants.

Planning for a channel through Joliet would require more study and dealing with the I&M Canal Commissioners and the City of Joliet. Complicating the matter was the physical setting north of Joliet. The river channel meandered from east to west across the valley floor, the I&M Canal and two railroads were on the east side of the valley, and several industries were along the river, some of which included water power developments. Route selection and design for the channel languished for several years (see map 13).

Route Selection and Decisions

Some on the board favored a costly navigable channel and water power development in this reach, and others favored a more modest approach to discharge the water from the Main Channel at Lockport. In May 1892, while the first set of bids were waiting to be opened for construction of the channel between Willow Springs and Joliet, the board ordered Chief Engineer Williams to prepare estimates for a tailrace channel to convey the water discharged from the Main Channel at Lockport if Proposition 2 were selected in the bidding process. Propositions 1, 2, and 3 are described in chapter 5. In its order, the board outlined some criteria for the tailrace, defined three sections of work for the 5.1 mile reach, and identified schemes for dealing with the railroads and the I&M Canal. The estimates that Williams developed were used in his analysis of the bids for Propositions 1, 2, and 3 but were not pursued further for decisions on the channel route from Lockport through Joliet.

In March 1893, Williams submitted a comprehensive and lengthy report on the Lockport to Joliet issue. By this time, construction of the Main Channel from Lockport to Summit was under contract, ending with Section 14, about a mile north of Lockport. Williams's report addressed two issues: the extension of the Main Channel to a point upstream of Lockport Road, which became Section 15 and the Lockport Controlling Works, and the tailrace channel from Lockport to Dam No. 1 at Jackson Street in Joliet. For the tailrace, he outlined three alternatives and provided for a total flow of 21,000 cfs, the combined maximum for the Main Channel and the Des Plaines River.

All three alternatives provided for a navigable channel with locks, potential for water power development, and capacity for maximum flow. Williams favored a channel along the west side of the valley because it avoided the railroads and industrial establishments, cost the least of the three alternatives, was the easiest to construct on, and had available land at reasonable cost. His report was referred to the committee and was the subject of intense consideration the following month (see map 14).

Upon the committee recommendation, the board directed in April that plans and specifications be prepared for a modified alternative, based on the majority report of the committee and supplemental reports submitted by the chief engineer. Trustee Cooley voted *nay*, then switched to *yea*, possibly so that he could move for reconsideration, although this never occurred. However, Cooley did submit a lengthy minority report. The modified alternative included the Main Channel extension upstream of Lockport, regulating works, discharge channel downstream of Lockport, embankments, bridges, etc., for an estimated cost of $1,325,000.

Nearly half of the cost was in the channel extension and regulating works. Cooley's objection to the majority report was the selection of a plan that provided for eventual navigation and water power development as these added to the cost and to the length of the Main Channel. He proposed a more modest approach, which was estimated at $880,000. Further action was delayed, perhaps by the departure of Williams in June. In October, the board ordered that the attorney obtain the ROW for the modified plan and that the chief engineer estimate its probable cost.

Chief Engineer Randolph reviewed all the options for the terminus of the Main Channel beyond Section 14 and for a tailrace and channel to and through Joliet and reported to the board in February 1894, requesting further direction. He found significant increases in cost resulting from new surveys and soil borings and from what he viewed as deficiencies in the previous plans. For the work downstream of Lockport, he estimated $535,000, whereas Williams had estimated $204,000. Randolph made no recommendations on a change in route.

In June, based on additional input from Randolph, the board adopted a modified approach, now referred to as the central route, at an estimated cost of $1,230,000. The end of the Main Channel and the regulating works would be located just north of Lockport Road, and downstream of that would be a tailrace channel along the west side of the valley to the crossing of the Joliet and Eastern Railroads. The end of the Main Channel would allow for eventual extension for navigation and water power development at a later time. Just upstream of the Joliet and Eastern Railroad crossing, the new tailrace channel would flow over a weir dam into the Des Plaines River. Downstream of the Joliet and Eastern railroad crossing, no additional work was thought necessary.

The work downstream of Lockport through Joliet was judged to be a relatively simple undertaking, perhaps requiring only one or two construction seasons. This proved not to be the case. Work proceeded upstream of Lockport, with Section 15 being awarded in December 1894 and the Lockport Controlling Works in January 1896. No action on the work downstream of Lockport occurred until May 1897, when the board ordered the preparation of plans and specifications for bidding and the preparation of an ordinance for acquisition of the ROW.

The ordinance was adopted in June, as well as another order instructing the chief engineer "to increase the scope of the plans to develop the fullest amount of water power which can be made available." The dream of water power held firm. The reality of the I&M Canal was recognized with another order calling for a conference with the I&M Canal Commissioners to address issues affecting the confluence of the tailrace channel and the I&M Canal in Joliet. The first contract for excavation of the tailrace channel and construction of the weir dam was advertised in June. Bids were opened in August and rejected in September due to high bids.

Yet another resolution was adopted in September asserting the SDC's right to cross the I&M Canal in Joliet and to use or modify the Upper and Lower Basins and the dams creating them, Dam Nos. 1 and 2, and to make other modifications as needed through Joliet to a point near the Rock Island Railroad crossing south of Jefferson Street. The resolution further called for all work to be performed by the SDC at no cost to the I&M Canal Commissioners and to be performed during a period of slack navigation. The resolution was sent to the commissioners, and the work for the tailrace and weir dam was readvertised with the expectation of completion by October 1898. The chief engineer expressed reservation regarding the progress of ROW acquisition and uncertainty over potential for serious delay caused by high water on the Des Plaines River. Keeping the river out of the work area was, unlike upstream of Lockport, not possible (see map 15).

Further delay came in November 1897 with the board's decision to return the bids for the tailrace and weir dam to the bidders unopened. The board also accepted offers to sell by two industrial properties along the modified route. This change of route significantly reduced the estimated cost of construction. This action was followed by adoption of an order for the removal of Dams No. 1 and 2 by dynamite or other means after the close of the navigation season on or about November 20, 1897, and calling for the SDC Police Department to patrol the area of the dams to secure the safety of human life and the protection of property.

The board had not heard back from the I&M Canal Commissioners responding to the resolution of September, and the board felt that the lack of response signaled acceptance. However, in December, the SDC heard from the City of Joliet, indicating that the SDC's plans were not acceptable to the city. The city required the rerouting of sewers and the raising of the grade of the Cass, Jackson, and Jefferson Streets to properly allow crossing of the new river channel. An agreement with the Rock Island Railroad was approved, providing for the ROW of the enlarged river channel, an additional span for the bridge, and the raising of the grade of the tracks.

Award of Contracts

Randolph's request for authority to advertise for bids for the tailrace in January was sent to committee and approved in February 1898 with added authority for the work to be designated as Sections 16 and 17. Section 16 would include the tailrace to just upstream of the Joliet and Eastern Railroad crossing, and Section 17 included enlargement of the Des Plaines River to the head of the Upper Basin near Ruby Street in Joliet. The board expressed a desire to proceed to advertise the work downstream of the Upper Basin, later designated as Section 18; however, plans were not available, and the board desired to meet with the I&M Canal Commissioners as soon as possible. After a few meetings with the Canal Commissioners, an agreement was drawn that was approved by both parties in March.

The agreement defined the physical setting of the I&M Canal through the Upper and Lower Basins and other relevant details and specified that the SDC would bear all construction costs, the I&M Canal Commissioners would provide their ROW to the SDC at no cost, and disputes would be resolved through arbitration. The SDC would widen and deepen the basins to accommodate 25,000 cfs, rebuild both dams, repair locks, and build new guide walls and a tow-path bridge across the river channel. Existing water power developments by private interests would be maintained. All work under the agreement would be done by contract except for a wall that separated the Middle Basin from the I&M Canal. Because of the priority of this wall to maintain navigation while work under contract was proceeding, the board ordered the chief engineer to negotiate an agreement to complete the wall expeditiously.

Bids for Sections 16 and 17 were opened in April 1898 and awards made in the same month. Section 16 was awarded as advertised, but Section 17 was modified, prompting a division of the board over the question of readvertising. The agreement with the I&M Canal Commissioners was executed after advertising, requiring some change in the contract. A majority of the board wanted to proceed with the work and not readvertise, thereby saving time, and an award was made. Following the award, the board ordered that the work downstream of Section 17 proceed expeditiously. Bids on Section 18 were opened in June, and an award was made in July. After six years of deliberations, all river channel work from Lockport through Joliet was finally under contract.

The work under Section 16 consisted of the excavation of the tailrace channel from the Lockport Controlling Works southward along the west side of the valley bottom for 14,100 feet, or 2.7 miles, to a point upstream of the Joliet and Eastern Railroad crossing on the natural river channel. In addition, it included a long embankment to create the east bank of the tailrace channel and embankments for the Lockport and Wire Mills (Division Street) Roads (see map 14).

Section 17 began just downstream of the Joliet and Eastern Railroad crossing and continued southward along the river to Dam No. 1, a distance of 7,600 feet or 1.4 miles. Work consisted of enlargement of the river channel, deepening of the Upper Basin downstream of the confluence of the river channel and the construction of levees and tow paths, including the Tow Path Bridge. No road or rail crossings were in this section (see map 15).

The work on Section 18 extended for 1.0 mile almost to McDonough Street along the river and consisted of excavation of the river channel; removal of Dams Nos. 1 and 2 and the Adams Dam; construction of a new Dam No. 1, embankments, and retaining walls; and the repair of Lock No. 5. Crossings in this section included Cass and Jefferson Streets and the Rock Island Railroad.

Canal Commissioners' Lawsuit

Following the agreement in March 1898, the I&M Canal Commissioners requested transfer of some property to them that had been acquired by the

SDC at the east side of Dam No. 1. This land was adjacent to the site of a water power plant, and the Canal Commissioners wanted the property to better utilize the water power to which they claimed a superior right over the SDC. Further, the commissioners wanted the SDC to pay for the land as it was claimed that the SDC's use of eminent domain to acquire the property was flawed. The board refused the request, and the commissioners went to the Circuit Court of Will County for relief.

The court order handed down in October 1898 was not favorable to the SDC and was taken up to the Illinois Supreme Court on appeal. Realizing that the appeal would not change the outcome, a settlement was worked out and approved by the Will County Court in December 1898 after the appeal was withdrawn. The court order declared the earlier agreement null and void, found that the right to use the water for power was vested in the state and not in the SDC, ordered the SDC to pay $20,000 to the Canal Commissioners, and required the SDC to carry out a number of construction details expanding on the earlier agreement. In essence, the earlier agreement, which the court found to have no statutory authority, was supplemented and converted into a court order.

A significant change in the work in Sections 17 and 18 occurred as a result of the settlement and court order. Chief Engineer Randolph, using the court order as authority, directed the contractors to carry out a number of contract changes in January 1900 without seeking board authority. Further, Randolph dictated the cost of the changes in his direction, indicating that due to the critical need to get on with the work, he was offering liberal prices in the face of the risk that the contractors were accepting to complete the work by the deadlines imposed. Much of the added work was for the purpose of maintaining navigation along the I&M Canal, but other items, such as additional river channel excavation and raising the height of levees, were to the benefit of added channel discharge capacity and flood protection for Joliet.

Water Power

The continuing dream of water power development became an objective with the passage of a board order in March 1898 calling for the preparation of "plans for the maximum development of water power incidental to the construction of the Drainage Channel at or downstream of the site of the controlling works upstream of Lockport." As the plans for work progressed, the only site suitable for water power development was the rebuilding of Dam No. 1 in Joliet. The board was of the opinion that the SDC had the right to the power available in the additional water brought from Lake Michigan.

Another board order adopted in June 1898 called for immediate advertising of "bids for the development and leasing of the water power incident to and susceptible of development along the Main Drainage Channel and its auxiliary channels or outlets, and subject to be controlled and disposed of by the" SDC. In the discussion before passage of this order, opinion was expressed that it was illegal for private interests to bid. The discussion also reflected public criticism that the City of Joliet would not be able to bid.

In August, just prior to opening the bids, the board placed limitations on the acceptability of proposals, requiring no interference with navigation, providing the right of the SDC or federal government to make changes to physical works for the benefit of navigation, and allowing the SDC to terminate the water power lease after ten years with the payment of damages for loss of investment if the power were needed for municipal lighting in Chicago. Two bids were received, one from a Joliet utility and the other from a Chicago businessperson.

The bids were subsequently withdrawn. The Joliet utility requested return of the bid deposit in September and the Chicagoan in November, both citing the unreasonableness of the limitations. The court order described above resolved the board's assumption of rights to the power available in the additional water. A legislative fix was required, which was eventually obtained. Nothing further was undertaken regarding water power until the 1903 legislation and construction of the Main Channel Extension.

Additional Agreements and Authorizations

Authorization of the work through Joliet was contained in an ordinance passed in July 1898 by the Common Council of the City of Joliet. The ordinance included the grant of ROW and the requirements for excavation of the river channel, removal and replacement of bridges at Cass and Jefferson Streets, plan review and approval by the City of Joliet, temporary bridges, sewer rerouting, and an indemnification. In return, the board adopted an ordinance in August accepting all the terms and conditions of the City of Joliet ordinance. The SDC assumed nearly all the costs for the improvements, and the City of Joliet agreed to pay for one half of the cost of the new Cass Street Bridge.

Three agreements were entered into that provided for some of the ROW. One agreement with the Santa Fe Railroad was an exchange of land near the Upper Basin for other land along the river channel route. The SDC not only received land for the river channel route but also land for the disposal of spoil near to the channel. The Santa Fe Railroad received land for improved routing of their tracks. The other two agreements provided ROW to the SDC for the channel in exchange for water, which was used for power by two industries. The SDC had the option of furnishing the equivalent amount of power in the form of electrical energy.

The Economy Light and Power Company owned and operated a generating station at Dam No. 1, and the SDC entered into an agreement to pay for the loss of water while the channel and new Dam No. 1 were under construction. The company bore the cost of the removal and replacement of the turbines and generators as they would, in the long run, benefit from the additional water available. The loss of water was limited to a period of eighteen months, and a bonus or penalty was provided if the period was less or more than eighteen months.

Contract Requirements

The three construction contracts were similar in the content of general conditions as the contracts for work upstream of Lockport. However, the format was changed, and details were added due to the structural work on walls and dams. Drawings were made a part of the contract, showing plans, cross sections, dimensions, grades, and some specifications. Less detail was written into the verbiage of the contract. The limits and location of the work were shown on the drawings and was not spelled out in the contract language. Specific reference was made to SDC ordinances recognizing the eight-hour workday, giving preference to union labor and providing for a $5 per hour penalty for each employee required to work more than eight hours per day.

In the contract for Section 16, three drawings were made a part of the contract. The work on the tailrace was defined in the specifications in six topics identified by roman numerals. Part I described channeling machine work west of the Lockport Controlling Works to make the vertical cut to 28 feet below CCD located 15 feet west of the gate abutments and piers. This cut isolated the Lockport Controlling Works from damage caused by blasting for excavation downstream. Part II defined the excavation to create the tailrace channel throughout the reach of the section. The grade of the channel bottom downstream of the Lockport Controlling Works was 18 feet below CCD.

Part III detailed the construction of an embankment from the Lockport Controlling Works to Wire Mills Road on the east side of the tailrace to confine the channel and block off low areas of the natural river channel. The embankment consisted of a compacted earth core flanked by rock fill. Part IV provided for the disposal of excavated material not used in the embankment in the low areas west of the Lockport Controlling Works. Part V required the removal of a previously constructed levee to protect work areas from flooding. Part VI detailed the embankments for Lockport and Wire Mills Roads. All material excavated was classified the same and paid for by the unit bid price. Completion was required by December 15, 1898.

Three drawings were made a part of the contract for Section 17. The work in this section was also defined in six topics identified with roman numerals. Part I detailed the construction of levees to divert and control the river to keep the work areas dry. Two levees were required to protect the work area and the I&M Canal from the river. Part II defined the pumping plant necessary to keep the work areas dry. Two steam-driven pumps rated at 20 cfs each at a head of 20 feet were specified in addition to the boilers and a ten-day supply of coal. Part III described the excavation of canal banks and the north and south parts of the Upper Basin. The canal bank removal was needed first to divert the river to the east. After construction of the west levee, the north part of the Upper Basin could be enlarged, followed by the deepening of the south part of the Upper Basin.

Part IV provided that construction of levees would consume all excavated material but defined other areas for extra material. Under Part V, special provisions gave the chief engineer wide latitude to direct changes in the field, required the contractor to work day and night when weather and water level allowed, required the contractor to be liable for all repair for flood damage, and provided that the SDC would pay for fuel and labor to pump out flooded excavation areas. Part VI required the use of channeling machines to make cuts to protect bridge foundations at SDC cost. For payment, materials were classified as earth, solid rock, or levee fill. Completion was required by May 1, 1899, with extensions given for interruptions due to flooding. Some of the work items in the contract were subsequently changed as a result of the court order previously described.

The contract for Section 18 included six drawings, and the work was defined under ten topics in the specifications. Part I defined the excavation of the Des Plaines River throughout the contract limits. Prior to the excavation of rock, the use of channeling machines was required along the upper reach of the west side to protect the east wall of the I&M Canal and around each bridge pier to protect the bridge foundation. Part II called for excavation of sediments in the I&M Canal, if required. This work could be accomplished either in the dry, when the canal was dewatered for lock reconstruction, or underwater after the work on the lock was complete. Part III required that suitable material for the core of the embankment must come from areas designated on the plans and must be compacted on suitable foundation material so as to be watertight. The remainder of embankment fill may come from the river excavation.

Part IV defined the removal of Dams Nos. 1 and 2 and the Adams Dam. Part V detailed the construction of the new dam for the preparation of foundations, mortar-faced concrete masonry, granite paving-block masonry, cut granite masonry, and the method of construction. Details were shown on the plans. The work on the dam was to proceed from east to west in 15-foot increments, each increment to be completed before starting the next. The dam was essentially a concrete monolith built in 6-inch lifts with granite facing.

Part VI defined the construction of a retaining wall along the west side of the river from Lock No. 5 to Jefferson Street. This wall formed part of the east wall of the I&M Canal. It was also a concrete monolith built in 6-inch lifts with Portland cement mortar facing. Part VII defined the repair of Lock No. 5 at Jackson Street at the west end of the new dam. Decayed and crumbling lock wall material was to be removed, and the voids filled with concrete and faced with mortar. The top of the lock walls were raised 2 feet using concrete with mortar facing.

In Part VIII, an inlet upstream of Dam No. 1 on the west bank and conduit was specified to provide for a connection to an existing bypass of Lock No. 5. Water was taken from the I&M Canal downstream of the lock and used for power by industries. The work consisted of a 5-foot diameter cast-iron pipe, a double-disc gate valve, and a protective concrete-masonry wall surrounding the pipe 50 feet in length. Part IX indicated that all excavated material not used in the embankment be deposited in areas identified on the plans. Part X defined miscellaneous work and required the construction of all cofferdams and protective levees at the contractor's cost, provided that low flows in the river would be handled by the SDC via the I&M Canal. The contractor was responsible for handling all other flows.

Next, the specifications set forth the sequence of construction as follows: (1) completion of the embankment core; (2) completion of the river channel and embankment; (3) completion of the retaining wall, except the part adjacent to the lock; (4) completion of the new dam; (5) cease work until the contractors on Section 17 have completed their work and the river flow discharges on the east side of the Upper Basin and Dam No. 1 and through the completed river excavation in Section 18; (6) construct cofferdams across the Upper Basin upstream of the north limit of the bypass west of the lock, across the south limit of the bypass, and across the area downstream of the dam where the uncompleted retaining wall is located; (7) complete the new dam, lock repair, retaining wall, and the bypass; and (8) remove the cofferdams and other obstructions. The actual sequence did not follow this path smoothly.

The specifications contained quality requirements for all the materials and about twenty unit prices. Excavation had only one classification but had three unit prices for river channel, I&M Canal, and core-material excavation. The work was to be completed for inspection before July 1, 1899. Some of the work items in the contract were subsequently changed as a result of the court order previously described.

Bridges

Despite the delay in beginning river channel excavation for the Lockport to Joliet reach, construction of the Joliet and Eastern Railroad Bridge was one of the first bridges completed, not only for this reach, but for all work between Chicago and Joliet. Work began in 1896 and was completed in 1897. Because of the high embankment on which the tracks crossed the river, this bridge was fixed. Construction of other bridges for this reach did not start until 1899. The Lockport Road and Wire Mills Road fixed bridges were begun and completed in 1899. Construction of the Rock Island Railroad, Cass Street, Jefferson Street, and Towpath Bridges were begun, but not finished, in 1899. Substructure work, however, was completed, which allowed for the increased flow in the river channel in January 1900. All the bridges in the Lockport to Joliet reach were fixed because these river channels were not considered navigable.

1898 Construction Season

Section 16 excavation and embankment work began in May and continued steadily to the end of the year. Temporary trestles for Lockport and Wire Mills Roads were begun in November. At the end of the year, work on the section was 59 percent completed. In Section 17, expedited work began in late March on the crib wall separating the I&M Canal from the Upper and Middle Basins and continued to completion at the end of May. Other work began in May, with construction of the protective levee to the west of the river and cutting through the towpath. A pontoon bridge was installed to maintain the towpath for navigation. Installation of the pumping plants allowed excavation of the new river channel to proceed west of the levee. Scrapers were used for earth excavation, whereas steam drills, blasting, and steam shovels were used on the rock. Rock removal was accomplished by inclines. The section was 29 percent complete at the end of the year. Work on Section 18 began in August with construction of embankment core and fill on the east side of the river, working from south to north. River channel excavation in rock began in November, also on the east side. Completion stood at 13 percent at the end of the year.

1899 Construction Season

Work in Section 16 for the first part of the year involved substructures for the Lockport and Wire Mills Road Bridges and embankments. Tailrace channel excavation resumed in June and was completed by December. In October, the embankment forming the east bank of the tailrace channel south of the Lockport Controlling Works was closed, forcing the low flow of the Des Plaines River down the tailrace and cutting off flow to the natural channel downstream of Lockport. Work on the section was 76 percent completed by the end of the year, reflecting work to be finished on the two roads.

Additional work was added to the contract for Section 17 by order of the chief engineer in January, but work early in the year was limited to the new channel north of Ruby Street and west of the I&M Canal and Upper Basin. Cold weather and spring floods caused delays. By midyear, work was well underway on the improvements to the I&M Canal and the dredging the Upper Basin. Upon completion of the Towpath Bridge foundations, the protective levee was removed in December, allowing the Des Plaines River to flow through the new channel.

The progress made by the contractor was not up to par, and several letters from the SDC were sent during the year calling attention to the progress requirements. The contractor was also cited for carelessness in performing blasting, causing a brief labor strike in October and not providing timely payment to laborers. Frequently the contractor was ordered to work weekends and round-the-clock shifts. By year's end, completion stood at 84 percent.

Except for the flood in March, channel excavation and construction of the east side embankment continued throughout the year in Section 18, starting the year south of Jefferson Street and progressing upstream. Improvements to the I&M Canal first focused on the replacement of Lock No. 5 and removal of the guard lock and the canal bypass so that these facilities would be ready for the navigation season. Next, efforts were devoted to the replacement of Dam No. 1, removal of Dam No. 2, and construction of the I&M Canal retaining walls and other channel walls. Temporary bridges and foundations for the several permanent bridges were completed before the end of the year.

Upon completion of Dam No. 1 in January 1900, the old dam was removed and the new channel was ready for the increased flow from Chicago. However, because of continued work on permanent bridges, maximum use was made of the I&M Canal for the Des Plaines River flow. The Section 18 contractor had to be reminded to make timely progress and was ordered to work weekends and round-the-clock shifts. At the close of 1899, 73 percent completion had been achieved.

References

Engineering Department Annual Report for 1899. Proceedings of the Board of Trustees of the SDC. SDC, 1900.

Engineering Works. SDC, August 1928.

SDC. Proceedings of the Board of Trustees of the SDC. 1893–1900.

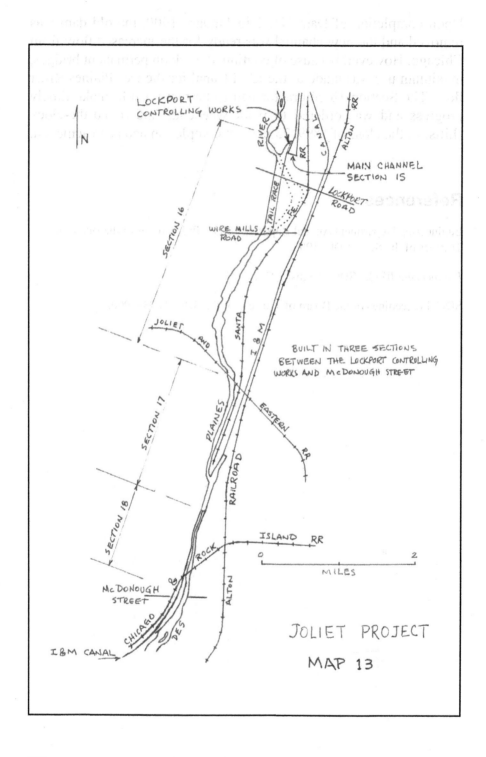

CHAPTER 8: JOLIET PROJECT

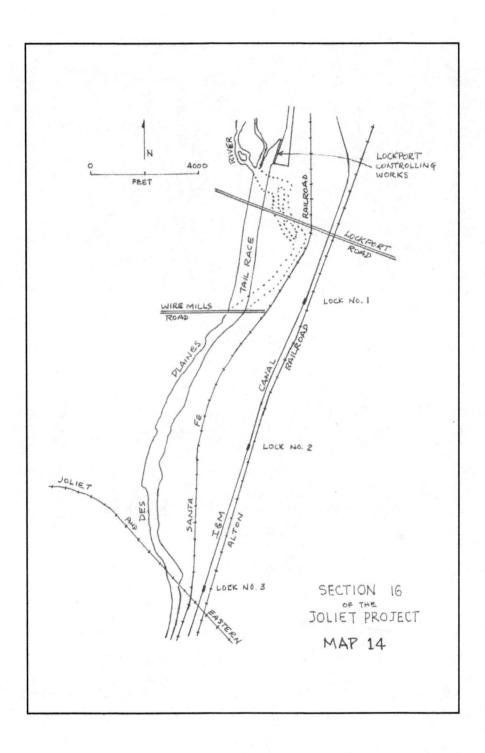

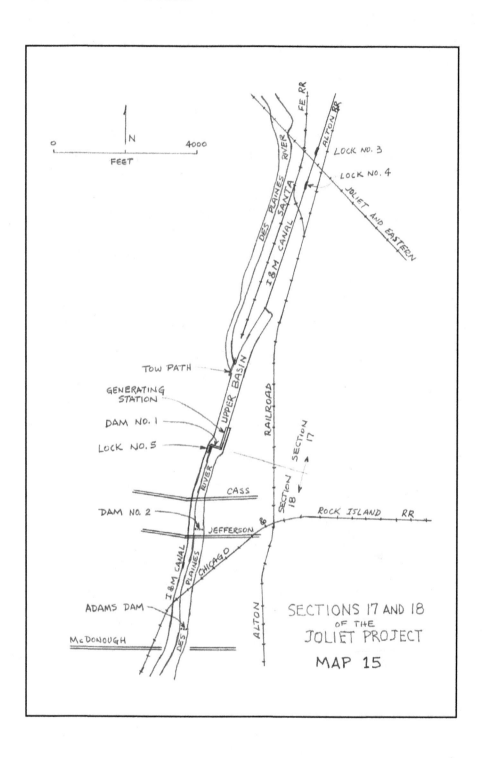

CHAPTER 8: JOLIET PROJECT

Photograph 8.1 taken on June 30, 1899 shows the excavation of the tailrace channel in Section 16. This channel was much broader and had less depth than the Main Channel upstream of Lockport. As can be seen, the valley floor is wide. The Des Plaines River was rerouted in the tailrace channel to the west side of the valley floor for the first mile downstream of the end of the Main Channel. MWRD photo, disc 9, image 10.

Photograph 8.2 taken on June 30, 1899 shows the wide and shallow tailrace channel at the Lockport Road Bridge. MWRD photo, disc 9, image 12.

Photograph 8.3 taken on April 26, 1899 shows the construction of the new Des Plaines River channel in Section 17 north of Joliet. This would be part of the Upper Basin, an in-channel reservoir upstream of Dam No. 1. MWRD photo, disc 12, image 1.

Photograph 8.4 taken on September 14, 1899 shows the excavation of the Des Plaines River channel in Section 17 looking north toward the Elgin, Joliet and Eastern Railroad crossing. The river channel was made wider through this reach. MWRD photo, disc 9, image 79.

CHAPTER 8: JOLIET PROJECT

Photograph 8.5 taken in late in 1897 shows the Des Plaines River on left and I&M Canal on right in Section 18. This view is looking south from the Cass Street Bridge with Jefferson Street Bridge and Guard Lock in the background. The Jefferson Street Bridge over the Des Plaines River is the original arch structure, soon to be replaced. MWRD photo, disc 4126, image 73.

Photograph 8.6 taken on September 18, 1899 shows the excavation of the new Des Plaines River channel through Joliet in Section 18 looking north at the Jefferson Street Bridge. The original bridge arch structure is behind the temporary timber and truss bridge. The I&M Canal is hardly noticeable to the extreme left behind the wall separating it from the river channel. MWRD photo, disc 13, image 5.

CHAPTER 8: JOLIET PROJECT

Photograph 8.7 taken on September 15, 1899 shows the excavation of the enlarged channel for the Des Plaines River in Section 18 looking north toward the Cass Street Bridge from the Jefferson Street Bridge. The I&M Canal parallels the river on the left behind the tree. The remnants of a low dam visible in the right foreground, was removed in this reach. MWRD photo, disc 13, image 6.

Photograph 8.8 taken on April 26, 1899 shows the I&M Canal in Section 18 looking north toward the Cass Street Bridge. The Des Plaines River is to the right and the wall between the canal and river protects the canal from variations in river level. This view is from near the Jefferson Street Bridge and the Guard Lock. MWRD photo, disc 11, image 95.

Photograph 8.9 taken on March 24, 1899 shows the reconstruction of the I&M Canal Lock No. 5 in Section 18. This view is looking upstream or north and the Des Plaines River is on the right. MWRD photo, disc 8, image 69.

Photograph 8.10 taken on March 24, 1899 shows construction of the tunnel used for emptying and filling Lock No. 5 on the I&M Canal in Section 18. Lock No. 5 and Dam No. 1 are adjacent to the Jackson Street Bridge. This view is looking south. The Des Plaines River is on the left. MWRD photo, disc 8, image 77.

Photograph 8.11 taken on March 24, 1899 shows work being performed to rebuilt Lock No. 5 on the I&M Canal in Section 18. This view is looking northerly. The vertical wall on the extreme right is a pier for the Jackson Street Bridge. MWRD photo, disc 8, image 76.

Photograph 8.12 another view taken on March 24, 1899 of the work being performed to rebuild Lock No. 5 on the I&M Canal in Section 18. The Des Plaines River and original Dam No. 1 is shown behind the workers with the original powerhouse on the far side of the dam. MWRD photo, disc 8, image 78.

Photograph 8.13 taken on April 26, 1899 shows the downstream or south end of the rebuilt Lock No. 5 on the I&M Canal in Section 18. The lock is immediately upstream of the Jackson Street Bridge shown in the top of the view. MWRD photo, disc 11, image 97.

Photograph 8.14 taken on April 26, 1899 shows a tow in the rebuilt Lock No. 5 on the I&M Canal in Section 18. The upstream lock miter gates are open, the downstream lock miter gates are closed and the tow in the lock appears to be headed north. The Des Plaines River is to the right. MWRD photo, disc 5, image 2.

Photograph 8.15 taken June 29, 1900 shows work in progress on the construction of a new powerhouse in Section 17. Looking north, the Jackson Street Bridge crosses in the foreground, the tailrace channel of the new powerhouse is seen in the center and the enlarged channel of the Des Plaines River is on the left. Construction of the new powerhouse began in 1899 and the enlarged river channel was completed prior to the end of 1899. MWRD photo, disc 15, image 42.

Photograph 8.16 taken June 29, 1900 shows the Jackson Street Bridge, the new powerhouse tailrace channel in the center foreground and the enlarged channel of the Des Plaines River beyond the bridge. In the left center under the bridge is seen the nappe of the new Dam No. 1 and turbulence of the Des Plaines River at the base of the dam. Jackson Street is the divide between Sections 17 and 18. The work upstream of the bridge is in Section 17. MWRD photo, disc 15, image 41.

Photograph 8.17 taken June 26, 1900 shows construction of the new powerhouse upstream of Jackson Street in Section 17. From the Jackson Street Bridge looking northerly, the tailrace channel is on the right. When completed and in operation, Des Plaines River water will exit the arched draft tubes along the base of the powerhouse. MWRD photo, disc 15, image 27.

Chapter 9

Chicago River Improvement

From the outset of their existence, the board was concerned about the condition of the Chicago River. Until October 1892, they explored ways to provide temporary relief for its frequently offensive and unsanitary nature. Because of the reason for which the SDC was formed, the board felt they must do something to relieve the river in the near term. Their searching was also prompted, no doubt, by the frequent criticism they received from the public. The construction of the Main Channel to provide permanent relief for the river was such an awesome undertaking and its completion seemed so far away, it must have been very tempting to do something in the immediate, even though it would accomplish little. But after late 1892, with the construction of the first 14 miles of Main Channel in progress and 6 more miles about to be put under contract, the board's attention shifted away from temporary relief for the river.

The Chicago River again became the focus of attention in August 1895. By this time, 28 miles of Main Channel were under contract, and rock excavation was nearly complete near Lemont. The board needed to improve the capacity of the South Branch of the Chicago River so that it would be able to deliver the flow of water from Lake Michigan required by the statute for the Main Channel. The South Branch had many bends and constrictive bridge openings, was shallow in spots, and was always busy with boat traffic. Fortunately, no time would be consumed in debate over the route (see map 16).

Work on the South Branch would also bring the SDC and the board face-to-face with the secretary of war because the US Army Corps of Engineers, acting under the secretary, maintained the Chicago River and Chicago Harbor and tried to exert its authority under new federal law. More on this interaction is explained later in this chapter and in chapter 11.

Temporary Relief

In a resolution adopted in December 1891, the board requested Chief Engineer Artingstall to estimate the cost of a survey of the Chicago River. His response the following month pointed to the lack of reliable information and the need for a comprehensive survey. He could not estimate the cost but thought that the effort would take two years. Next, the board sought clarification on their legal obligations in two areas. Was the SDC authorized to prevent the construction of sewers discharging to Lake Michigan? Can the SDC expend funds on temporary relief of the river by improvement of the I&M Canal?

In February 1892, Attorney Goodrich's response was *no* and *maybe*, respectively. In the first issue, the city and other municipalities had exclusive authority to provide for their own drainage. The SDC was authorized to remove pollution through the construction of channels, not to prevent further pollution by interfering with the powers granted to municipalities. On the second issue, such expenditures were authorized if the purpose was, either now or later, for permanent relief. However, if the expenditure was solely for temporary relief, not part of a permanent measure, then it was not authorized.

Seeking further clarification, the board requested an opinion on the use of the I&M Canal for temporary relief if it was to be the permanent channel. There was a change in attorneys, so it was Attorney Carter who now answered that the I&M Canal had, in fact, been used for temporary relief since 1848, and such use was recognized by the Illinois General Assembly. There was no demand in the statute requiring the SDC to provide relief, except through construction of a channel that conforms to the act. In an attempt to conclude the matter, an order was introduced, calling upon the chief engineer to determine if the I&M Canal or its ROW could be used to allow the SDC to relieve the

Chicago River prior to completion of the Main Channel. The order failed to be adopted.

Chicago notified the SDC in August 1892 of their intent to increase the capacity of the Bridgeport Pumping Station to 1,330 cfs, recognizing that this would reach the capacity of the I&M Canal but would not completely alleviate conditions of the river, especially during floods. Chief Engineer Williams was ordered by the board to make surveys and cost estimates of the diversion of flood waters from the North Branch at Grosse Point, now Skokie, to Lake Michigan at Winnetka. This diversion would relieve the Chicago River of excess flow and increase the effectiveness of the Bridgeport Pumping Station. The following month, Williams reported that, based on field reconnaissance, the diversion was feasible. Again, the board ordered the chief engineer to conduct surveys and examinations, prepare maps, and show routes of all prior diversion routes.

Meanwhile, another failed attempt was made in September 1892, following Shovel Day, for temporary relief using the I&M Canal. This attempt was an order for the chief engineer to determine how 2,500 cfs could be diverted from the South Branch to the I&M Canal or Des Plaines River, or both, prior to completion of the channel and not later than the summer of 1893 and, if a feasible plan was found, for the attorney to advise the board if the SDC could do this work itself or in conjunction with Chicago.

The attempt to "order" a solution was tabled. However, at the following meeting, a resolution was adopted that, after citing all previous orders and resolutions on the matter and the progress of work on the Main Channel, called for the board, the chief engineer, and the attorney to consider any practical plan to provide temporary relief via discharge to the Des Plaines River—even enlargement of the I&M Canal and pumping to the Des Plaines River at Summit, either in cooperation with Chicago or by the SDC alone.

A voluminous committee report with several appendices came out in October that thoroughly explained the history of the river's problem, population growth, public health effects, all facets of the remedy, pumping at Summit, legal issues, and conclusions. It ended with these recommendations: (1) Des Plaines River flood waters be eliminated from Chicago, (2) expedite the surveys for the North Branch diversion,

(3) increase the capacity of the I&M Canal to 2,000 cfs, (4) prepare plans for a 2,000 cfs pumping plant at Summit, (5) consider other measures prior to opening of the Main Channel, and (6) draft needed legislative authority for the 1893 session of the general assembly. The board adopted the report and the recommendations.

The appendices were not necessarily supportive, however. Chief Engineer Williams pointed out that numerous bridges would have to be raised, and the discharge of sewage-laden waters to the broad and shallow Des Plaines River would cause offensive odors and an unhealthy air. The secretary of the Illinois State Board of Health advised against discharge to the Des Plaines River due to the obviously ill effects. The Commissioner of the Department of Public Works indicated that Chicago would proceed to increase the capacity at the Bridgeport pumping plant to 1,300 cfs and no more unless others would expand the capacity of the I&M Canal. The general superintendent of the I&M Canal indicated that they had no authority to participate or funds to contribute toward use of the I&M Canal as proposed. Another appendix, unsigned, gave the history of similar attempts for temporary relief that never came to fruition.

As explained in chapter 7, it was about this time that the board adopted the I&M Canal as the Main Channel route between Bridgeport and Summit. Perhaps it was this decision, the end of the warm weather, preoccupation with putting other channel work under contract, or the negative implications in some of those appendices; but there was no further mention of the need for temporary relief. The following May, the board abandoned the I&M Canal as the Main Channel route due to legal problems with the I&M Canal Commissioners. Public criticism was again pointed at the SDC for offensive conditions in the Chicago River with the spring runoff in 1893. It was charged that excess flows from the SDC construction near Summit had caused the Des Plaines River to flow to Chicago, overwhelming the pumping to the I&M Canal at Bridgeport. Chief Engineer Randolph's June 1893 report demonstrated that Des Plaines River flooding was lower than in 1892, the contractor on Section F did not disturb the levee, and the Bridgeport pumps were operating below capacity.

Public criticism did not abate, however, and President Wenter delivered a message in August 1893 calling for the SDC to take over operation of the Bridgeport pumping plant. At his urging, a committee was formed to call upon the mayor to propose the transfer of the Bridgeport pumping plant to the SDC. The committee did not submit a formal report to the full board.

Plans for Improvement

The Chicago River, from its mouth on Lake Michigan to Robey Street on the South Branch, was known to have inadequate capacity to convey the flow intended for the new channel. In May 1894, the chief engineer was ordered to investigate and "formulate a scheme and cause plans to be prepared" for works to cleanse the South Branch and contribute to the health of its environs coincident with the construction of the Main Channel. Over a year later, in August 1895, surveys and a plan were completed for increasing the capacity of the South Branch to 5,000 cfs. The work would involve dredging to deepen, replacement of docks to widen, replacement of existing bridges with longer spans to eliminate restrictive bridge openings, and construction of the Bypass Channel from north of Jackson Boulevard to south of Van Buren Street to add river flow capacity.

The Bypass Channel was needed where bridge replacement was not feasible due to the existence of an underriver tunnel, bridge piers, and permanent structures. The cost was estimated at $873,000. Instead of proceeding to advertise the work, the board wanted a plan for the North Branch and South Fork as well, and the same was ordered to be prepared.

Chief Engineer Randolph was requested in December 1895 to prepare the plans for improvement of the Chicago River and South Branch in greater detail, but it was another year before authority was given to advertise the work in February 1897. Bids were received and an award made in May. In the same month, a resolution was adopted calling for the acquisition of the ROW between Van Buren and Robey Streets, noting that the plans require approval by the secretary of war. Although the contract called for dredging between the mouth of the river and Robey Street, the ROW and work actually began on the South Branch at Lake Street. Dredging the Chicago River to a 16-foot depth from its mouth to Lake Street was undertaken by the US Army Corps of Engineers in the late 1890s, so the SDC did not have to improve this reach of river.

It should have been obvious to Chicagoans that progress was being made now that work would be underway right in the heart of the city. Nevertheless, public criticism again arose, this time questioning the authority of the SDC to improve the river as planned. In response, the board ordered the attorney in July to prepare an opinion on the question. However, the opinion was never rendered formally. In support of the SDC, the Chicago River Improvement Association adopted a resolution endorsing the work to improve the river.

In September, Randolph presented a map and ordinance of the ROW needed for the river improvement. Although the ordinance was adopted, it was rescinded at the following meeting and another adopted in its place, which differed only slightly. Authority to acquire the ROW was given the following month, and at the same time, the use of condemnation for purchase of ROW between Adams and Van Buren Streets for the Bypass Channel was approved. An order authorizing a board committee to carry out negotiations and condemnation was adopted in December.

Due to the existence of an operating railroad in the vicinity of Adams and Van Buren Streets, purchase of the ROW authorized above was not pursued. The railroad objected to the purchase and threatened a long battle over condemnation. Instead, it was decided to enter into agreements for building the covered Bypass Channel on the west side of the river through a subterranean easement. This easement would allow the railroad to use the roof of the Bypass Channel, and the SDC would pay an annual fee for the easement. In February 1898, an agreement with the Pennsylvania Company, operator of the Pittsburg and Fort Wayne Railroad was executed, providing for a covered conduit 50 feet wide and a bottom elevation of -18 feet, CCD. The SDC was responsible for design, construction, and maintenance; and the railroad had to approve the plans. A nearly identical agreement was executed with the Alton Railroad in June but for a smaller area of the Bypass Channel.

In March, authority was granted to advertise for bids for the Bypass Channel. The bids were opened, and awards were made in June 1898. Separate contracts were awarded for the substructure and the superstructure.

Consideration was given for another bypass structure near Taylor Street in order to provide the requisite river flow capacity. However, as a result of a study of alternate plans for a bypass and for replacement of the Taylor Street and Chicago Terminal Railroad center-pier bridges with bascule bridges and the result of discussions with the city and the Chicago Terminal Railroad, it was found that replacement of the bridges was less costly. The board approved the recommendation of the committee in November 1898. The new bridges were of the Scherzer design, and the SDC agreed to bear the cost of removal of the center-pier bridges and the construction of the new bridges. Bridge replacement would not only provide the needed river flow capacity but would also remove the center piers, which were obstructions to navigation. These two new bridges were built on the original river channel, which was eventually relocated in straightening the South Branch in 1928, at which time the railroad bridge was relocated to the new river channel.

Something that perhaps was not known or was overlooked at the time of the agreement with the Pennsylvania Company and when the Bypass Channel substructure contract was awarded was the need in March 1899 to enter into a two-party agreement with the West Side Elevated Railroad and a three-party agreement with the Pennsylvania Company and the West Side Elevated Railroad. These agreements were needed because of a water intake and conduit for the West Side Elevated Railroad, which ran through the Bypass Channel. Water was withdrawn from the river at the rate of 44 cfs by the West Side Elevated Railroad for an electrical generating plant located at Van Buren and Throop Streets. The intake and conduit within the Bypass Channel ran between Jackson and Van Buren Streets. The SDC agreed to remove the existing intake and conduit and to build a new intake in the west wall of the Bypass Channel. The multiple agreements, executed in June, were necessary to provide for the transfer of rights, access, and maintenance.

Permits from the Secretary of War

Up until the work in the Chicago River, the SDC did not have to obtain permits from any other government agency for their construction activities. Even the work in the Des Plaines River was not under any other regulatory authority. The SDC was simply carrying out the

Illinois statutory authority for the construction of channels to reverse the flow of the Chicago River. But now, in the South Branch, the work came under the authority of the secretary of war granted by the US Congress in the Rivers and Harbors Act of 1890. The South Branch was considered a navigable waterway, and this act required that any work in navigable rivers must be permitted by the secretary of war. There was some contention between the SDC and the US Army major, the local representative of the secretary of war, who was overseeing the federal work on the Chicago Harbor and River. The contention arose over the federal priority for work on the Calumet River, rather than the Chicago River, and the repeated lack of response to requests for federal funding for part of the work in the construction of the Main Channel.

The first permit received was in November 1898 for construction of a cofferdam to the east side of the Adams Street Bridge center pier. This cofferdam was intended to protect construction of the Bypass Channel. The next two permits were received in January and March 1899 for the Chicago Terminal Railroad and Taylor Street Bridges, respectively. In May, a permit was obtained for construction of the cofferdams for the Bypass Channel between Jackson Boulevard and Van Buren Street. Nearly one year after construction on the Bypass Channel had begun, the permit was obtained. Obviously, the SDC was not intimidated by the secretary of war or federal authority. It was not until July 1900 that the permit for the dredging of the South Branch was obtained, over three years after the contract for the work was awarded and work began.

Another permit was received in May 1899 for opening the Main Channel and connecting it to the Chicago River. This permit was granted under new authority in the Rivers and Harbors Act of March 1899, which required the secretary of war to permit any excavation, filling or altering the course or capacity of all navigable rivers, harbors, and channels.

Contract Requirements

The contract for dredging described the work as being between the mouth of the river and Robey Street, but the ROW acquired started at Lake Street. Bottom grade was -20 feet CCD without any slope from Lake to Robey Streets. Approximately 600,000 cubic yards was to be dredged at locations shown on the plans or as directed by the chief engineer. Disposal was at locations provided by the contractor. Disposal actually occurred in Lake Michigan or Lake Front Park, now Grant Park. The contractor was required to comply with all ordinances of the city and regulations imposed by the US Government. The order of the work was at the discretion of the chief engineer, except that work in the West Fork from Ashland Avenue to Robey Street, which was specifically mentioned to be the first to be accomplished. The stated prices were for excavation by the cubic yard, for dock removal by the lineal foot, and for pile removal per pile. Payment was at monthly intervals. Construction of new docks was handled under extra work provisions. All work was required to be completed by November 1, 1898.

Both of the contracts for the Bypass Channel substructure and superstructure were similar in format and contained detailed specifications for materials, shop drawing submittal and approval procedures, and even specified that all drawings shall be 26 by 40 inches in size. The general conditions were similar in both contracts. At the time of award, the board added a liquidated damages penalty of $100 per day for exceeding the time of completion requirement.

The contract for the Bypass Channel substructure included three drawings and specified several work items as follows: clearance of the work site; construction of cofferdams to keep the river out of the work site; installation of piling to retain the sidewalls of the excavation; installation of a foundation for the retaining wall that would be determined by the chief engineer after the excavation was open, and would be either a timber grillage or wooden piles driven into the subsoil; the retaining wall to be installed could be either monolithic concrete or brick masonry; installation of portable bridges and temporary timber trestles for railroad and street crossings over the work site; sewers and water mains encountered were to be rerouted as directed by the chief engineer; installation and operation of pumps to keep the work site dry; dismantling a portion of a freight house and re-erecting the freight house upon completion of the Bypass Channel; truncating the corner of an office building and rebuilding the exterior and interior walls; and the restoration of all roadway paving. Payment was based on measurement by the chief engineer and eighteen unit prices were provided based on the bidding. All work was to be completed by December 15, 1898, which was 5.5 months after the execution of the contract.

The work to be performed in the superstructure contract was detailed on several drawings. Material specifications adopted from the Pennsylvania Company were included for the structural steel and other metals used. Other specifications for sand, stone, cement, mortar, concrete, painting, and shipping were given. The contract was priced on a lump sum basis for all work. Unit prices for concrete and steel were included to allow for a change in quantity. Rather than the monthly progress payments customary in other contracts, this contract provided for payment of 60 percent upon delivery of the materials to the work site and the remaining 40 percent upon completion, subject to the normal retainage of 12.5 percent. All work was to be completed by February 1, 1899, which was seven months after the execution of the contract and forty-seven days following the completion requirement for the substructure. The penalty clause added by the board at the time of award was stricken in December 1898 upon the request of the contractor.

Bridges

Although there were numerous bridges across the South Branch, the improvement of the river did not require the construction of new bridges. The only bridges built were the two bascule bridges to replace the center-pier bridges for the Chicago Terminal Railroad and Taylor Street. As explained above, the bridges were replaced to remove obstructions to navigation and to improve river flow capacity. In addition, a new approach span was built for the Van Buren Street crossing over the Bypass Channel. The contract for the Van Buren Street approach span was awarded in December 1898. Contracts for the Taylor Street and Chicago Terminal Railroad Bridges were awarded in May 1899. At the end of 1899, these bridges were 43, 39, and 23 percent completed, respectively. An agreement was executed with Chicago for the construction of a new bridge at Canal Street in August 1899, but no work was performed as of the end of 1899.

1898 Construction Season

Although the dredging contract was awarded in May 1897, lack of ROW held up the beginning of work until late in the year. By the start of 1898 only 4 percent of the dredging work had been completed. Dredging resumed in March and continued throughout the year, working from south to north in the South Branch between bridges. Chief Engineer Randolph determined that dredging in the bridge openings should be accomplished after the dredging in the river between bridges was completed. Removal and/or replacement of docks was performed along with the dredging. Work on the Bypass Channel substructure began in July with the construction of cofferdams in the river and the driving of piles for temporary trestles. Work continued until early November, when failure of the cofferdam caused flooding of the work site and cessation of work until late December when repair of the cofferdam and dewatering were complete. Overall, work on the Chicago River Improvement was judged to be 46 percent complete by the end of the year.

1899 Construction Season

Dredging was suspended for two months due to extremely cold weather early in the year. After resumption in March, it continued until suspended again in mid-September due to a dispute with Chicago over deposition of dredge spoil in the lake. The SDC agreed to deposition in Lake Front Park and granted additional cost to the contractor for the extra handling. Dredging resumed in October and was completed by December in the reaches of the river between bridges. Dredging in the bridge openings began in December. At the end of the year, dredging work was 89 percent complete.

Work on the Bypass Channel continued through the cold weather, and concrete placement for the walls began in March. The driving of piles near Jackson Boulevard in May caused settlement of the swing bridge such that the bridge would not close properly. However, this settlement did not delay work on the Bypass Channel, which now was proceeding night and day, seven days per week. Structural steel for the superstructure was delivered in March, but erection could not begin until July. The upstream end of the Bypass Channel was completed in October and flooded in November. A temporary bulkhead allowed work to continue on the lower end. Repairs were made to nearby buildings and other structures whose settlement was attributed to the excavation for the Bypass Channel. At year's end, work on the Bypass Channel was 71 percent complete.

References

Hill, Libby. *The Chicago River: A Natural and Unnatural History*. Chicago: Lake Claremont Press, 2000.

SDC. Engineering Department Annual Report for 1899. Proceedings of the Board of Trustees of the SDC. 1900.

SDC. Proceedings of the Board of Trustees of the SDC. 1893–1900.

CHAPTER 9: CHICAGO RIVER IMPROVEMENT

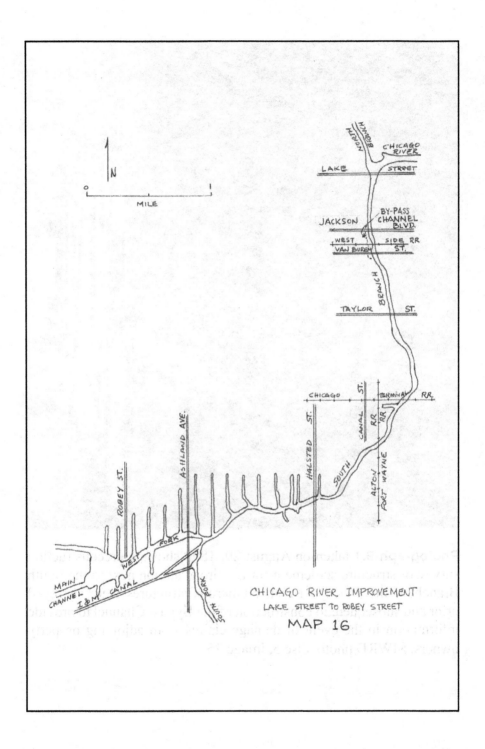

CHICAGO RIVER IMPROVEMENT
LAKE STREET TO ROBEY STREET
MAP 16

Photograph 9.1 taken on August 29, 1898 shows a preconstruction survey of structure settlement for a railroad station next to the South Branch of the Chicago River. Numerous structures were surveyed prior and subsequent to construction of the Bypass Channel to provide information in the event of damage claims from adjoining property owners. MWRD photo, disc 5, image 26.

CHAPTER 9: CHICAGO RIVER IMPROVEMENT

Photograph 9. 2 taken on August 31, 1898 shows work underway in construction of the Bypass Channel. Looking north from under the Jackson Boulevard Bridge, the South Branch is to the right. The beams running from left to right form the roof of the Bypass Channel and the upstream entrance to the channel is just beyond the track-mounted derrick. MWRD photo, disc 5, image 34.

Photograph 9.3 taken on March 8, 1899 looking north from south of the Jackson Boulevard Bridge shows work in progress on construction of the Bypass Channel. The center pier of the bridge is in the center of the view and the Bypass Channel is beneath the beams crossing the view and passes to the left of the center pier. MWRD photo, disc 8, image 64.

CHAPTER 9: CHICAGO RIVER IMPROVEMENT

Photograph 9.4 taken on August 17, 1899 shows construction of the Bypass Channel under the West Side Elevated Railroad Bridge next to a railroad warehouse. The railroad bridge seen in view consists of two trusses west of the west bascule bridge leaf and pier. The two trusses span the Bypass Channel and ground level railroad tracks and terminal buildings. MWRD photo, disc 9, image 43.

237

Photograph 9.5 taken August 17, 1899 shows construction of the Bypass Channel. The west pier and foundation of the bascule span of the West Side Elevated Railroad Bridge is seen in the center and the Bypass Channel is to the left of the pier. The South Branch is to the right. The elevated railroad crossed the South Branch between Jackson Boulevard and Van Buren Street. Behind the railroad bridge is seen the Jackson Boulevard Bridge, a center pier swing bridge. MWRD photo, disc 9, image 45.

CHAPTER 9: CHICAGO RIVER IMPROVEMENT

Photograph 9.6 taken October 5, 1899 shows placement of the steel beams forming the roof of the Bypass Channel. View is looking north toward the Adams Street Bridge and the upstream entrance to the Bypass Channel. MWRD photo, disc 13, image 27.

Photograph 9.7 taken October 5, 1899 shows the inside of the river wall cofferdam of the Bypass Channel. The dipper dredge was used to excavate sediment in the South Branch along the river wall and in front of the entrance and exit of the Bypass Channel. The underside of the West Side Elevated Railroad Bridge is seen to the left behind the dredge. MWRD photo, disc 13, image 29.

CHAPTER 9: CHICAGO RIVER IMPROVEMENT

Photograph 9.8 taken October 5, 1899 shows a full view of the dipper dredge. The West Side Elevated Railroad Bridge is in the left background. MWRD photo, disc 13, image 31.

Photograph 9.9 taken April 6, 1900 in the Bypass Channel looking north from Van Buren Street. Use of the channel was not initially needed to effect the reversal of the river because flows were far below the maximum. MWRD photo, disc 14, image 22.

CHAPTER 9: CHICAGO RIVER IMPROVEMENT

Photograph 9.10 taken April 6, 1900 in the Bypass Channel looking south from the entrance north of Jackson Boulevard. Remnants of the structure are visible in 2011 along the west bank of the South Branch north of the Van Buren Street Bridge, next to the Union Station tracks. MWRD photo, disc 14, image 19.

243

Photograph 9.11 taken on April 27, 1904 shows the upstream entrance to the Bypass Channel on the South Branch. The entrance in below the railing and steel girder in the lower left quarter of the view. View is looking southerly from Adams Street. The Jackson Boulevard Bridge center pier is between the river channel and the Bypass Channel. Beyond Jackson Boulevard is the two truss spans of the West Side Elevated Railroad Bridge. The double leaf bascule railroad bridge is to the left of the truss spans. Eventual replacement of center pier swing bridges and bascule bridges with wider river clearance would eliminate the need for the Bypass Channel. MWRD photo, disc 25, image 30.

Chapter 10

Bridges

The act made reference in many locations to a navigable waterway, to maximum velocities, and to minimum depths and widths, all of which defined the conditions for the safe passage of commercial watercraft. However, the act made no reference to bridges. The construction of bridges over the Main Channel was necessary to allow land travel on roads and railroads. Lacking statutory definition made the matter of bridges discretionary to the SDC and its board. It was the cause of much debate and division among the members of the board. It may appear that the construction of bridges was an afterthought, given that the many bridges were put under contract late in the 1890s and several were not completed when the Main Channel was placed in service in January 1900. Such was not the case. The subject was in the background from the beginning.

The SDC built or funded the construction of thirty-one bridges to effect the reversal of the flow of the Chicago River. Actually, considering that the Eight-Track Bridge includes four leaves, it can be said that the SDC built thirty-four bridges. This does not include the many temporary bridges necessary to allow traffic to pass while construction of the permanent bridge was underway.

Early Considerations

Deliberations on the choice of route and size of channel during 1890 and 1891 did not consider the type or size of bridges. Between Lockport and Bridgeport, there were seven locations where railroads were crossed and six locations for roads regardless of the channel route. Consultations with the railroad and road owners were not mentioned in the proceedings, but there had to be interest and inquiries. The first action was a resolution adopted in December 1891, requiring that all permanent bridges over the Main Channel be movable, either swing or draw, and not fixed or stationary. The resolution also directed the chief engineer to contact railroad officials to establish arrangements for crossing their ROW. Thus, the first policy on bridges called for movable bridges. Property owners along Canal Street in Chicago approached the board at this time requesting that a proposed crossing be built using a drawbridge to allow unimpeded passage of navigation.

In January 1892, after the shake-up in the board explained in chapter 4, former chief engineer and now board member Lyman E. Cooley criticized the lack of planning, citing the estimates based on fixed or stationary bridges, and failure to contact the railroads. He concluded that the navigable requirement demanded proper railway and highway crossings. In the urgency to begin excavation of the Main Channel, none of the twenty-nine contracts awarded for excavation included bridge construction. These contracts did provide, however, that the contractor must allow the SDC or other contractors to construct bridges within the work limits of the excavation contracts. In order for the excavation work to proceed to completion and maintain road and railroad crossings, the excavation contractors were given extra compensation to construct temporary timber trestle and girder bridges. These served until the permanent bridge was in place and in service.

Indecision Over the Type of Bridges

With rock so close to the surface west of Summit, subsurface exploration was not necessary for bridge foundations. The depth of rock was not known generally east of Summit, so in March 1894, authorization was given for soil borings at all bridge sites. Meanwhile, design of bridges for Lemont, Romeoville, and Willow Springs Roads was underway. In November, when bids were opened on alternative designs, disagreement broke out over whether these bridges should be swing or fixed. In majority and minority reports, the debate raged between whether to build the lower cost fixed bridge now and a more costly movable bridge later versus building the higher cost movable bridge now. Although the former was the majority position and the bridge policy was changed to fixed bridges, it was decided to delay a final decision on these three bridges because they need not be constructed until the Main Channel was nearing completion.

The following June, a similar debate occurred over a railroad bridge. In this case, an agreement with the railroad was at stake, and the urgency was to obtain approval to excavate the Main Channel across the railroad ROW. Following majority and minority reports, the decision was to agree to build a fixed bridge with a provision that at a later time, the bridge could be converted to a swing bridge.

A consultant named William Hughes approached the SDC in April 1896, suggesting that he be retained to prepare designs, plans, and specifications for all bridges. The board rejected this offer and ordered that such work be performed by the SDC under the direction of the chief engineer.

The difficulty in reaching agreement on the issue of fixed versus movable bridges led to a legislative solution. Without statutory authority, there would eventually have been a cloud over the approval of completion of the Main Channel and approval by the governor because the fixed bridges could be viewed as a contravention of the navigable waterway requirement. Further, the railroads must have been nervous over the future of this uncertainty. The act was amended effective July 1, 1897, requiring that all bridges be of a type that would allow bridges to be operated as movable bridges no later than seven years after the date when water is turned into the Main Channel.

As the legislation was in progress, the board expressed its desire to move forward. In April 1897, the board ordered the chief engineer to prepare the most economical and practical designs for swing bridges over the Main Channel and submit all contracts for railroad bridges for advertisement. This order was amended in June to include design and contracts for substructures. This change from a fixed bridge policy to a movable bridge policy rendered unusable all the previous design work on fixed bridges, no doubt a frustration for the Engineering Department. Further modification of the bridge design approach occurred in August, with the board directing that all advertisements for bridge construction contracts be based on designs prepared by the SDC unless the bidder can offer a competitive design that was superior in the opinion of the chief engineer.

Progress in putting bridge construction under contract proved elusive. By the close of 1897, only five contracts were awarded for bridge construction, and only one bridge was completed. The railroads raised objections to some designs and specifications, and revisions were needed. To forestall contractors from using unsuitable masonry products, the board ordered that limestone from the Main Channel excavation in the Des Plaines River valley be used in all bridge abutments. Cement for mortar and concrete was purchased by the SDC and supplied to contractors. A separate Bridge Department was organized under the chief engineer. Additional orders by the board were issued for the preparation of contracts for advertisement.

The use of center-pier bridges was more economical than draw-type bridges for the width of channel to be crossed. Further, the center pier facilitated the construction of a fixed bridge that could more easily be converted to a movable swing-type bridge as required in the act. Nevertheless, navigation interests induced the board to consider a resolution in July 1898 to abstain from construction of center-pier bridges due to the federal navigation policy and the desire for the Main Channel to accommodate interstate commerce. This resolution was not adopted, but bridges over the South Branch and the first railroad bridge on the Main Channel were built without center piers. All other bridges between Robey Street and Willow Springs were built with center piers.

Delay over the question of center-pier bridges brought a communication in November 1898 from the contractor whose work was being held up on the substructure for the Chicago Terminal Railroad Bridge over the South Branch and the Romeoville Road Bridge over the Main Channel. The matter was referred to committee. With the potential for litigation, extra cost, and bad public image, the board was moved to decide that the bridge over the South Branch would be built as a single-leaf bascule while the Romeoville Road bridge would be built as a swing bridge, with the pier off the channel on the west wall. While older swing bridges were restrictive to navigation and water flow in the South Branch, the newer swing bridges would not be the same over the wider Main Channel.

Work on bridges began to move forward. From the past, William Hughes resurfaced in April 1898 to be engaged as a consultant to design the Belt Line Railroad Bridge in Section K. A year later in May 1899, Hughes was put on the payroll as engineer of bridges and bridge construction, reporting to the board's Committee of Engineering.

Swing Bridges

Once the act was amended, the preferred bridge design was the center-pier swing type. Bridges over the Main Channel north of Willow Springs were of this type, except for the Eight-Track Bridge. With the center pier, the overall weight of the bridge was less than a bascule bridge. Less weight translated to less cost. Essentially, the bridge consisted of two truss sections spanning from the center pier to the abutment on each channel bank. Until the bridge had to be made movable—that is, to swing—the two trusses could be independent of each other since the center pier and the abutments supported each end of a truss. Upon being made movable, the two trusses were tied together with a structural frame over the center pier, allowing the two spans to be supported as cantilevers when the bridge was swung off the abutments. To make these bridges movable, the top of the center pier was fitted with a circular track, and the underside of the bridge structure above the center pier acted as a large turntable.

In total, eight center-pier swing bridges were constructed, three for roadways and five for railroads. The roadways were Southwestern Boulevard/Western Avenue, Kedzie Avenue, and Summit-Lyons Road. The railroads were Madison and Northern Railroad, Santa Fe Railroad Twenty-Sixth Street Line, Belt Line Railroad, Santa Fe Railroad Main Line, and Chicago Terminal Railroad. The three road bridges have been replaced. The Belt Line Railroad Bridge was not completed when the Main Channel was opened in January 1900. The foundation work was complete, and trains continued to use the temporary timber trestles. The permanent bridge was completed in 1901. Four of the five railroad bridges remain in service in 2011, although they are no longer movable. The Chicago Terminal Railroad Bridge has been modified due to a railroad accident, and it is no longer movable. All bridges have adequate clearance for navigation.

Downstream of Willow Springs, where the Main Channel is a vertical wall section 162 feet wide, the bridges were the bobtail swing type. The pier was located on the north or west channel wall. The truss over the Main Channel was longer to span from the pier to the abutment on the opposite wall. The opposite truss was shorter and counterweighted. Four bobtail swing bridges were constructed. Three were for roadways and one for the Santa Fe Railroad Main Line. The roadways were Willow Springs Road, Lemont Road (Stephens Street), and Romeoville Road. The railroad bridge remains in service in 2011, although it is no longer movable. All three road bridges have been replaced with high-level fixed spans. However, the Romeoville Road swing bridge has been saved as a landmark and was relocated on the Centennial Trail, just north of Romeoville Road between the Main Channel and the Des Plaines River.

Fixed Bridges

Several fixed bridges were necessary adjuncts to facilitate the Main Channel construction. Included were the Madison and Northern Railroad viaduct over Kedzie Avenue in Section O and the Santa Fe Railroad viaduct over Stephens Street. These two bridges remain in service in 2011. In addition, several road and railroad bridges of the fixed type were built over the Des Plaines River between Summit and Joliet. The railroad bridges remain in service whereas the road bridges have been replaced.

The Eight-Track Bridge

Perhaps the most challenging of the bridges was the design and construction of the Eight-Track Railroad Bridge located at 2500 West (Campbell Avenue) over the Main Channel in Section O. Three railroads occupied the same ROW at this crossing: the Pittsburg and St. Louis, the Stock Yards, and the Northern Pacific. Collectively, they required eight tracks. Early negotiations on an agreement for access to excavate the Main Channel resulted in the decision to build the Collateral Channel as described in chapter 7. The railroads also obtained the right to approve the plans for the bridges to be built by the SDC. The first agreement was executed in June 1895, and the last in November 1895.

The excavation contractor built timber trestles for the temporary girder bridges, and these were offset on detours to allow space for construction of the permanent bridge. The agreements were modified in April 1897 to accommodate the change in the act. The original design was based on a center-pier swing bridge, and contracts were advertised, bids taken, and contracts awarded by February 1898. Having believed that the railroads were in agreement with the swing bridge design, it came as a shock to the SDC in March 1898 when the railroads announced their preference for a drawbridge design. The board annulled the just-awarded contracts immediately.

Not wanting to lose any more time, the board also ordered the advertisement of design proposals acceptable to the railroads. Two respondents were C. L. Strobel and the Scherzer Rolling Lift Bridge Co., and each was awarded a fee for the preparation of designs used for contractors to bid on. Bids were opened in June and reported to the board in August. Other designs had been submitted by some bidders, so the bid analysis and railroad design approval took extra time. Following presentation to the board in August, there was divided opinion, but the construction bid submitted by Scherzer based on the Scherzer design was approved and a contract was awarded.

One of the unsuccessful bidders petitioned the court for an injunction barring the SDC from executing the contract. The court's restraining order was issued, and the board directed in October that the contract be readvertised to avoid delay while the litigation over the contract was in process in the advent of an adverse decision. In December, the court's decision upheld the contract award, and the restraining order was dissolved. This occurred just before the new bids were to be opened, and all bids were returned unopened. Meanwhile, the excavation contractor was given extra work to rework some of the temporary bridge trestles to allow channel excavation to be completed, which was originally intended to be completed with the permanent bridge construction. The delay in the permanent bridge contract award would have pushed the channel excavation completion into the next century.

The Eight-Track Bridge was a remarkable design for the time. The bascule bridge was becoming popular in urban areas where there was a need for an unobstructed crossing of a navigable channel. At this crossing, eight tracks were needed for the three railroads. A swing bridge would have been less expensive, but the width would require a center pier of large diameter, causing the span to be longer. The swing time was of concern to the railroads as it would have been longer than the raising and lowering time of the bascule. The crossing was accomplished with four double-track, single-leaf bascule bridges. The four bascule bridges were in close parallel arrangement, with alternating hinges on the north and south banks. In December 1899, as the SDC was preparing to open the Main Channel, work on the permanent bridge was well underway, but completion would extend beyond the December 31, 1899, due date in the agreements with the railroads. An extension of time was granted to the contractor.

Upon opening the Main Channel in January 1900, the foundation work for this bridge was complete, and trains were still using the temporary timber trestles. Four interim fixed-span trusses were installed in 1900 so the timber trestles could be removed and boats could pass with limited clearance. The four bascule spans were completed in 1902. Three of the four bascule spans remain in service in 2011, although they are no longer movable. One span has been replaced due to age and corrosion.

Santa Fe System

Six bridges had to be built to accommodate crossing the Santa Fe Railroad's ROW to reverse the flow of the Chicago River. Three swing bridges were built over the Main Channel in Sections N, H, and 8. Three fixed bridges were built, two over the Des Plaines River diversion channel in Sections F and 8 and another over the Stephens Street viaduct in Section 8. In addition, railroad embankment realignment and regrading was necessary in Sections H and 8. Most of this work was provided for in one agreement, often referred to as the Santa Fe System Agreement. Other dealings with the Santa Fe Railroad involved a transfer of land parcels in Joliet to facilitate the I&M Canal realignment and temporary use of their tracks in Chicago to reroute traffic of other railroads necessary to stage the construction of the Eight-Track Bridge.

South Branch Bridges

Only two bridges over the South Branch were put under contract before the end of 1899. Due to federal control of navigable waters, both required permits from the secretary of war, and both were required to be constructed as single-leaf bascule bridges. These were the Taylor Street and Chicago Terminal Railroad Bridges. The substructures were complete, or nearly so, at the end of 1899. The superstructures, however, were constructed early in the next century. One additional bridge was built, the Van Buren Street fixed span across the Bypass Channel. The Taylor Street Bridge was removed in the late 1920s when the South Branch was straightened. The Van Buren Street span was replaced with the construction of a new bascule bridge. The Chicago Terminal Railroad Bridge was relocated in 1928 with the straightening of the South Branch.

Joliet Project Bridges

In this reach, only the I&M Canal was navigable, and no bridges were required to be built over it by the SDC. All bridges built spanned the Des Plaines River and were of the fixed type built of steel girders or trusses spanning between masonry piers. Seven bridges were constructed by the SDC. These include five road bridges for Lockport Road, Wire Mills Road (Sixteenth or Division Street), I&M Canal Towpath, Cass Street, and Jefferson Street and two bridges for the Joliet and Eastern and the Rock Island Railroads. The latter railroad bridge remains in service in 2011, and the former was replaced with the waterway project in the 1930s. All other bridges were removed or replaced with the Illinois Waterway Project construction in the 1930s.

Other Bridges

All the bridges mentioned above were necessary to complete the projects that resulted in the reversal of the flow in the Chicago River and South Branch. In the following years the SDC made commitments to build fifteen more bridges over the Chicago River and South Branch to enlarge the waterway cross-sectional area and reduce hazardous currents and flow conditions.

References

Illinois Laws. Illinois, 1897.

SDC. Proceedings of the Board of Trustees of the SDC. 1890–1900.

CHAPTER 10: BRIDGES

Photograph 10.1 taken in 1895 shows quarried rock from the excavation of the Main Channel near Lemont that will be used for masonry walls and bridge abutments and piers. MWRD photo, Geiger set, image 144.

Photograph 10.2 taken on June 16, 1900 shows fabricated steel delivered to the construction site of the Belt Line Railroad Bridge over the Main Channel in Section K. The superstructure of some of the bridges was not completed until after the river was reversed. MWRD photo, disc 15, image 25.

CHAPTER 10: BRIDGES

Photograph 10.3 taken on June 27, 1899 show the temporary trestle construction to route active railroad tracks around the construction site of a permanent bridge. This is the crossing for three rail lines that eventually became the Eight-Track Bridge over the Main Channel in Section O. MWRD photo, disc 9, image 25.

Photograph 10.4 taken on April 28, 1899 shows the temporary trestle constructed for the Belt Line or Western Illinois Railroad Bridge over the Main Channel in Section K. The temporary trestle is not yet completed in the view and the train is crossing the original tracks on original ground not yet excavated for the Main Channel. This bridge was not completed until 1901, but the foundations, center pier and abutments were completed by the end of 1899. MWRD photo, disc 12, image 2.

CHAPTER 10: BRIDGES

Photograph 10.5 taken on September 7, 1898 shows a pile foundation for a railroad bridge abutment in the Earth or the Earth and Rock Section where bedrock was not close enough to the surface for the foundation to be built on rock. Hence, wooden piles were driven into the glacial till. MWRD photo, disc 5, image 43.

Photograph 10.6 taken on September 7, 1898 shows the deep water-filled excavation for a railroad bridge center pier foundation in the center, the pile foundation for the abutment in the foreground and the temporary timber trestle to the right. MWRD photo, disc 5, image 44.

Photograph 10.7 taken on September 7, 1898 shows work in progress to build the masonry center pier and one abutment for a bridge over the Main Channel in the Earth or Earth and Rock Section. MWRD photo, disc 5, image 42.

Photograph 10.8 taken on September 7, 1898 shows the completed masonry abutment for a bridge over the Main Channel. MWRD photo, disc 5, image 39.

CHAPTER 10: BRIDGES

Photograph 10.9 taken on September 11, 1899 shows the excavation for an abutment for the Eight-Track Railroad Bridge over the Main Channel in Section O. The temporary timber trestle crossing is in the background and in the distant background beyond the trestle is the top of the center pier tower for the Western Avenue/Southwestern Boulevard Bridge. MWRD photo, disc 9, image 73.

Photograph 10.10 taken on September 11, 1899 shows a wide view of foundation work for the Eight-Track Bridge over the Main Channel in Section O. Temporary timber trestles were built on each side of the work area to route the three rail lines using this crossing around the construction site. This was another bridge not completed until late 1900. MWRD photo, disc 9, image 75.

CHAPTER 10: BRIDGES

Photograph 10.11 taken on September 11, 1899 shows the laying of masonry stone on top of a concrete cap on wooden piles for the north pier of the Eight-Track Railroad Bridge in Section O. This pier will support a fixed truss spanning from the north abutment and one end of the four bascule spans across the Main Channel. MWRD photo, disc 9, image 72.

Photograph 10.12 taken on September 14, 1898 show the preparation of the bedrock surface for the foundation of the swing pier for the Santa Fe Railroad Bridge over the Main Channel in Section 8. This view, taken from the embankment for the temporary crossing, is looking upstream with the pier located on the northwest side of the channel. Towers for cableways for removing rock from the channel excavation are seen in the left background. MWRD photo, disc 5, image 55.

Photograph 10.13 taken on December 8, 1898 shows the installation of the rollers for the swing bridge turntable for the Santa Fe Railroad in Section 8. This view, looking southwesterly, also shows the rock wall left in the Main Channel excavation to support the temporary timber trestle for the railroad, and beyond is the top of the temporary timber trestle for the Lemont Road (Stephen Street) crossing. A cableway tower remains in the left upper background. MWRD photo, disc 11, image 5.

Photograph 10.14 taken on February 16, 1899 shows the erection of the superstructure for the Lemont Road Swing Bridge over the Main Channel in Section 8. The temporary timber trestles for the crossings for the Santa Fe Railroad and Lemont Road are also shown, the former in the foreground and the latter in the left background. MWRD photo, disc 8, image 21.

Photograph 10.15 taken on March 7, 1899 shows a test of the swing for the Lemont Road Swing Bridge in Section 8. The temporary trestle for the Santa Fe Railroad and railroad embankment rock is in the foreground. Part of a cableway tower is seen to the extreme left. MWRD photo, disc 8, image 56.

Photograph 10.16 taken on February 6, 1899 shows the beginning of erection of the superstructure for a center pier swing bridge. MWRD photo, disc 7, image 96.

CHAPTER 10: BRIDGES

Photograph 10.17 taken on February 24, 1899 shows the erection of a center pier swing bridge, most likely the Santa Fe Railroad in Section N. MWRD photo, disc 8, image 35.

Photograph 10.18 taken on February 24, 1899 shows further erection of the superstructure for a center pier swing bridge. The temporary bridge is to the extreme left. MWRD photo, disc 8, image 36.

Photograph 10.19 taken on September 8, 1898 shows erection of the superstructure of a center pier swing bridge in the Earth Section. MWRD photo, disc 5, image 47.

Photograph 10.20 taken on September 8, 1898 shows erection of the center tower for a center pier swing bridge. These types of bridges are essentially two truss bridges connected by the center tower. MWRD photo, disc 5, image 50.

CHAPTER 10: BRIDGES

Photograph 10.21 taken December 12, 1898 shows the partially completed superstructure for the Chicago, Madison and Northern Railroad Center Pier Bridge in Section O. The excavation of the soil left for construction at this crossing is being removed. MWRD photo, disc 11, image 15.

Photograph 10.22 taken on September 14, 1898 shows the erection of the superstructure of the Santa Fe Railroad center pier swing bridge over the Main Channel in Section G. MWRD photo, disc 5, image 61.

CHAPTER 10: BRIDGES

Photograph 10.23 taken October 11, 1898 shows the completed Chicago Terminal Railroad Center Pier Swing Bridge over the Main Channel in Section E. The temporary timber trestle has not yet been removed. MWRD photo, disc 6, image 1.

Photograph 10.24 taken November 15, 1898 shows the placement of steel girders atop masonry piers for the Santa Fe Railroad crossing of the Des Plaines River adjacent to Section 8. The temporary timber trestle crossing is seen to the left. MWRD photo, disc 10, image 84.

CHAPTER 10: BRIDGES

Photograph 10.25 taken April 13, 1899 shows the raising of the truss spans for the Chicago Terminal Railroad Bridge over the Des Plaines River adjacent to Section E. This bridge didn't have to be replaced, but had to be raised to accommodate raising the grade of the rail line for the crossing of the Main Channel. One through truss was added as shown in Photograph 10.26. MWRD photo, disc 11, image 84.

Photograph 10.26 taken on May 2, 1899 shows the collapse of the raised spans for the Chicago Terminal Railroad crossing of the Des Plaines River due to shifting false work supporting the spans. MWRD photo, disc 12, image 16.

Photograph 10.27 taken on September 15, 1899 shows the raised truss spans for the Chicago Terminal Railroad crossing of the Des Plaines River. The masonry work on three of the raised piers has been completed and work is in progress on the nearest pier on the right. A through truss has been added on the left to enlarge the opening under the bridge for increased flow in the river. MWRD photo, disc 13, image 2.

Photograph 10.28 taken on August 22, 1899 shows the final grading of the road bed approaching a new bridge over the Main Channel. MWRD photo, disc 9, image 58.

Photograph 10.29 taken on October 18, 1899 shows the construction of the Stephens Street viaduct under the Santa Fe Railroad adjacent to Section 8. The timber trestle under the railroad cars will be removed and replaced with steel girders spanning between the concrete abutments. The timber piles will be cut below grade, the remaining fill removed and paving for Stephen Street completed. The railroad swing bridge spanning the Main Channel is in the center background. MWRD photo, disc 13, image 48.

Photograph 10.30 taken on September 11, 1899 shows the completed Southwest Boulevard/Western Avenue Bridge over the Main Channel. This is a four lane, center pier, swing bridge. Water is in the Main Channel at this point because this is connected to the South Branch. The Main Channel is blocked by a fill across the channel at the Eight-Track Bridge construction site. This view is from the temporary trestle of the Eight-Track Bridge crossing, looking northeasterly. MWRD photo, disc 9, image 76.

CHAPTER 10: BRIDGES

Photograph 10.31 taken on August 26, 1901 shows the interim Eight-Track Railroad Bridge crossing the Main Channel in Section O. The four through truss spans between piers in the channel will be replaced with four bascule spans for the permanent bridge. This interim arrangement provided for the Main Channel excavation to be completed and railroad traffic to be unimpeded. The design of the bascule spans was not awarded until late in 1899. The four bascule spans were constructed in succession to minimize railroad operation interruptions. MWRD photo, disc 17, image 19.

Photograph 10.32 taken on January 10, 1899 shows the completed Chicago, Madison and Northern Railroad Bridge in Section N. This is a two track, center pier, swing bridge. Work is in progress to complete the excavation of the Main Channel. This view is from the southerly bank, looking northeasterly. MWRD photo, disc 4126, image 98.

Photograph 10.33 taken on August 22, 1899 shows the completed Kedzie Avenue Bridge over the Main Channel. This is a two lane, center pier, swing bridge. This view is from the southerly bank, looking westerly. MWRD photo, disc 9, image 51.

Photograph 10.34 taken on June 28, 1900 shows the completed Santa Fe Railroad Bridge in Section N. This bridge is a two track, center pier, swing bridge. Water has been flowing in the Main Channel, the Chicago Sanitary and Ship Canal, since January 17, 1900. This view is from the southerly bank, looking westerly. MWRD photo, disc 15, image 33.

CHAPTER 10: BRIDGES

Photograph 10.35 taken on December 29, 1900 shows the completed Belt Line or Chicago & Western Illinois Railway Bridge in Section K. This is one of the bridges completed after the Main Channel was placed in service. It is a four track, center pier, swing bridge, with two tracks within the through truss and one track on each side outside the through truss. MWRD photo, disc 16, image 92.

Photograph 10.36 taken on September 27, 1898 shows the completed Santa Fe Railroad Bridge in Section G. This is a two track, center pier, swing bridge. This view is from the southerly bank looking westerly. MWRD photo, disc 5, image 86.

Photograph 10.37 taken on June 28, 1899 shows the completed Summit-Lyons Road Bridge in Section F. This is a two lane, center pier, swing bridge. The view is from the top of the northwest bank of the Main Channel downstream of the bridge. Excavation of the Main Channel is nearing completion in this portion of Section F. Beneath the left span of the bridge is seen the reduction in size of the Earth Section of the Main Channel. MWRD photo, disc 9, image 2.

Photograph 10.38 taken on April 13, 1899 shows the completed Chicago Terminal Railroad Bridge in Section E. This is a two track, center pier, swing bridge. This view is from the top of the northwest bank looking northeasterly. MWRD photo, disc 11, image 87.

Photograph 10.39 taken on May 14, 1900 shows the completed Willow Springs Road Bridge. This is a two lane, bobtail, swing bridge. The view is from the top of the southeast canal wall of the transition from trapezoidal to rectangular cross-section of the Main Channel, looking southwesterly or downstream. Water has been flowing in the Chicago sanitary and Ship Canal since January 17, 1900. MWRD photo, disc 13, image 87.

Photograph 10.40 taken on April 3, 1900 shows the completed Santa Fe Railroad Bridge in Section 8. This is a two track, bobtail, swing bridge. This view is taken from the northwest side of the Main Channel looking in a southerly direction. Water has been flowing in the Chicago Sanitary and Ship Canal since January 17, 1900. MWRD photo, disc 14, image 24.

CHAPTER 10: BRIDGES

Photograph 10.41 taken on July 6, 1900 shows the completed Lemont Road (Stephen Street) Bridge. This is a two lane, bobtail, swing bridge. This view is from the Santa Fe Railroad Bridge looking southwesterly in the downstream direction. Water has been flowing in the Chicago Sanitary and Ship Canal since January 17, 1900. MWRD photo, disc 15, image 58.

Photograph 10.42 taken on September 6, 1899 shows the temporary timber trestle bridge built for the Western Stone Company at the division between Section 9 and 10. The company required rail access from the company stone quarry west of the Main Channel to the I&M Canal east of the Main Channel. The company used the I&M Canal to transport its quarried rock to points upstream and downstream. This trestle was removed prior to the end of 1899. MWRD photo, disc 9, image 63.

Photograph 10.43 taken on August 8, 1899 shows the completed Romeoville Road Bridge and the temporary timber trestle. The permanent structure is a two lane, bobtail, swing bridge. At the east end of the bridge is a ramp enclosed in masonry walls to accommodate two right angle turns in the road to cross the Santa Fe Railroad tracks that are close to the Main Channel. MWRD photo, disc 9, image 38.

BUILDING THE CANAL TO SAVE CHICAGO

Photograph 10.44 taken on April 13, 1899 shows the completed Santa Fe Railroad Bridge over the Des Plaines River adjacent to Section F. This is a two track, six span, fixed girder, bridge. This view is taken from the downstream side of the bridge on the west bank of the river looking in a northerly direction. MWRD photo, disc 11, image 75.

CHAPTER 10: BRIDGES

Photograph 10.45 taken on April 13, 1899 shows the completed Summit-Lyons Road Bridge over the Des Plaines River adjacent to Section F. The new bridge is a two lane, through truss on new piers. The original five fixed truss segments were incorporated into the new crossing on rebuilt piers. MWRD photo, disc 11, image 78.

Photograph 10.46 taken on September 15, 1899 shows the nearly completed Chicago Terminal Railroad Bridge over the Des Plaines River adjacent to Section E. The new bridge is a two lane, through truss. The original three fixed truss spans were incorporated into the new crossing on rebuilt piers. See Photographs 10.25, 10.26 and 10.27 for additional detail. MWRD photo, disc 13, image 1.

CHAPTER 10: BRIDGES

Photograph 10.47 taken on March 1, 1899 shows the Santa Fe Railroad Bridge over the Des Plaines River adjacent to Section 8. This is a two track, fixed plate girder, 12 span, bridge. This view is from the southeast river bank descending bank looking northerly. MWRD photo, disc 8, image 52.

Photograph 10.48 taken on January 27, 1899 shows the Lemont Road Bridge over the Des Plaines River adjacent to Section 8. This is two lane, fixed timber truss and beam, 10 span, bridge. This view is from the southeast river bank looking northerly. MWRD photo, disc 11, image 50.

Photograph 10.49 taken on February 27, 1900 shows the Lockport Road Bridge over the Des Plaines River in Section 16. This is a two lane, through truss, three span, bridge. This view is from the center of the road looking west. The wide expanse of water is the new Tail Race Channel downstream of the Lockport Controlling Works. MWRD photo, disc 14, image 33.

Photograph 10.50 taken on August 4, 1899 shows the Wire Mills Road Bridge over the Des Plaines River in Section 16. The new structure is a two lane, through truss, three span, bridge. The original bridge was a single span and this span was incorporated into the new bridge. More spans were needed to cross the Tail Race Channel. This view is from the east river bank looking southwesterly. MWRD photo, disc 9, image 23.

CHAPTER 10: BRIDGES

Photograph 10.51, undated, shows the Elgin, Joliet and Eastern Railroad Bridge over the Des Plaines River at the division between Sections 16 and 17. The new bridge consists of one through truss and four truss spans. The through truss provides for navigation clearance over the new river channel. These five spans were added to plate girder spans over railroad tracks and the I&M Canal. MWRD photo, disc 4126, image 67.

Photograph 10.52 taken on June 29, 1900 shows the Cass Street Bridge in Section 18. This is a two lane, through truss single span bridge over the Des Plaines River. The I&M Canal lies to the west of the river on the other side of the low wall. The canal is spanned by a plate girder bridge. This view is from the east bank looking southerly. MWRD photo, disc 15, image 39.

Photograph 10.53 taken on April 20, 1900 shows the Jefferson Street Bridge in Section 18. This is a two lane, through truss, two span, bridge over the Des Plaines River. The original bridge over the I&M Canal to the west remained in service. The temporary bridge is seen behind the new bridge. The original bridge was a multi-span masonry arch. MWRD photo, disc 14, image 6.

Photograph 10.54 taken on June 29, 1900 shows the Chicago, Rock Island and Pacific Railroad Bridge in Section 18. The new single span section of this bridge is a four track, fixed plate girder. The truss span in this view is the original bridge over the I&M Canal to the west of the river. This view is from the east bank looking southerly. MWRD photo, disc 15, image 37.

CHAPTER 10: BRIDGES

Photograph 10.55 taken on July 6, 1900 shows the Stephen Street viaduct under the Santa Fe Railroad adjacent to Section 8. This is a two track, fixed plate girder, single span, bridge. MWRD photo, disc 15, image 44.

Chapter 11

Ancillary Issues

Building the Main Channel was not all design and construction. Many other issues had to be dealt with by the Board of Trustees and their department heads, which arose out of the construction activity and relations with other public and private organizations. The SDC took on these issues and mostly dealt with them effectively so that the work to reverse the flow of the Chicago River was not impeded.

Worker Health

Although covered by the specifications, worker health was not foremost in the minds of the contractors. Each contractor had a camp built for the workers to place the workers in proximity with the work site. Commuting from home and family to the job did not exist on the project. Also, many workers were itinerants or immigrants and did not have family nearby or a place of residence. Unsanitary conditions in the camps brought complaints, and the state board of health communicated with the SDC in September 1892 on the need for monitoring of the camps and for the contractors' compliance with specifications. In October, the board adopted a plan to hire a sanitary inspector, to establish clinics along the construction route, and to adopt sanitary regulations for the camps. All the planned improvements were in place by December.

The board adopted rules for the sanitary inspector in January, and upon recommendation of the Chicago Board of Health, the sanitary inspector was made an officer of the state board of health in March 1893. Thus, the sanitary inspector had state authority, and it was not necessary for the SDC to seek this authority amended into the act. The sanitary regulations for the camps were amended to include per capita air volume and floor space requirements in May.

Workers without family in the Chicago area had no one to call upon to provide for a decent burial when the worker died. The contractors had no responsibility, even when the death occurred on the job. Thus, the SDC agreed to the sanitary inspector's recommendation to provide for a burial lot and a proper service. Burials were made in cemeteries near the construction sites.

The sanitary inspector submitted monthly reports to the board on conditions in the camps and on the incidence of communicable disease along the construction route. The occurrence of disease was rare, and the camps were kept in proper condition once regular inspections were conducted and penalties administered to contractors. In November 1895, the sanitary inspector resigned, and the board decided not to continue the position since the worker health and sanitation regulations were being administered by the SDC construction resident engineers.

Labor Relations

Aside from worker health and sanitation, other labor relations issues confronted the SDC, the first of which was the eight-hour workday. Although the workday was covered in the specifications, a question arose as to whether the SDC must enforce the requirement. In late 1892, the board deferred action as the matter was before the court. A decision was rendered in June 1893, placing responsibility for the enforcement on the entity issuing the contract. The board invited the carpenters' union to submit enforcement language for the contracts. The language with modifications was placed in the contracts by the summer of 1893. Simply put, any work beyond eight hours in a day required extra payment.

Some contractors attempted to pay workers in script or notes redeemable at contractor concessions. An end to this practice occurred in August 1893 by a resolution of the board. In February 1894, the minimum wage was established at fifteen cents per hour. Contractors were required to pay workers at semimonthly intervals, but this requirement was not always followed. The board adopted a policy in May 1895 to withhold contractor payments if workers were not paid as required. Also, contractors were required to pay discharged employees in cash.

Although labor unions had become accepted, it was not until 1897 that the board adopted an ordinance establishing a preference for union labor in certain contracts. This ordinance was debated, and its adoption was not unanimous. Pro and con factions of the board submitted majority and minority positions.

The Employed and the Unemployed

The construction of the Main Channel brought a paying job to many people. However, there were not enough jobs to go around to all who needed work. The board was sensitive to both these issues. The contractors' workforce was meticulously tracked during the first three years and reported to the board monthly. SDC field engineers would keep a weekly count of the employed for each contract. The average for the month was reported to the board in the monthly report of the Engineering Department. As shown in table 1, October 1894, with 5,728, was the month with peak contractor employment. By this time, work was underway on twenty-nine contracts between Chicago and Lockport. The reporting of the number employed was stopped without explanation in the report for January 1895. It is unlikely that the workforce was more than it was in October 1894 as construction activity was beginning to wind down and some of the contracts were completed by the summer of 1895.

Problems with the unemployed were brought to the attention of the board by labor unions. In an effort to help, day labor was hired to expedite work on the Des Plaines River diversion channel in 1893 and 1894. The work was administered as extra work under several contracts, and the day laborers were paid by the contractors.

Table 1: Average Number of Men Working Each Month

Month/Year	1892	1893	1894
January		907	2,522
February		676	2,448
March		588	3,175
April		726	3,778
May		1,419	4,207
June		1,331	4,860
July		2,113	5,100
August		2,948	5,682
September	189	4,032	5,033
October	690	3,543	5,728
November	985	2,924	5,286
December	1,046	2,441	5,234

Notes:

1. Laborers and workers were referred to as *men* since women were not considered as part of the workforce in the late nineteenth century.
2. Number of men working was taken from the monthly reports of construction activity as found in the Proceedings of the Board of Trustees.
3. Reporting the number of men working was discontinued in the monthly reports starting with the report for January 1985. No explanation was given for the discontinuance.
4. Months and number of contracts awarded are as follows: June 1892, fourteen; November 1892, two; February 1893, four; December 1893, six; May 1894, two; September 1894, one. All channel excavation and retaining wall work was under contract from Robey Street to Lockport.

Public Order and Police Power

The contractor's camps were not diverse populations but were mostly composed of homogeneous ethnic or national groupings. Rowdiness would often occur with occasional fights been the camps of differing nationalities. This disorder was intruding upon the quiet, out-of-the-way small towns and settlements along the construction route. Complaints made their way from the town leaders to the SDC Board and other officials. This disorder was also disruptive to construction progress, and contractors often took the law into their own hands in hiring guards to protect their workers. The board adopted a report on public order in October 1892 and received an opinion from the attorney the following January indicating that authority was available for the SDC to maintain public order on its property and to protect the construction sites.

The board communicated with and met with the towns along the construction route in February 1893. The SDC requested the towns to provide order through their police officers. However, the towns felt that it was beyond their means to provide police protection given the level of construction activity. Some towns had no regular police officers. The contractors also requested protection for their workers from violence caused by unemployed persons. The SDC agreed to seek statutory authorization for police power. The legislation was successful and was approved by the governor in June 1893. A police department was formed, a marshal appointed, and requirements for police officers adopted in July.

The first monthly police department report to the board came in October. These regularly featured a listing of crimes committed, persons arrested, dead bodies found, fights broken up, etc. Additional police officers were hired in February 1894, and stations were established along the construction route as additional work was put under contract. Regulations were expanded in the summer of 1894 to prohibit concealed weapons and saloons serving alcoholic beverages from SDC property.

In 1895 and later years, the SDC Police Department was an effective authority for preserving the peace in the Des Plaines River valley along the channel route. Unlicensed saloon keepers were the only recurring problem, and these were dealt with forcibly. There was also a problem with unauthorized use of SDC property, and the board gave the police the power to serve notice and move to evict the parties. As construction activity wound down in the late 1890s, the number of police officers was reduced, stations along the route closed, and unneeded equipment sold.

Floods and Water Supply

The construction of the Main Channel was to solve the problem of pollution of the water supply, and construction of the Des Plaines River diversion channel was to solve the problem of floods from the Des Plaines River reaching and damaging Chicago. As a result, the construction activity was of interest to the city and its population. Fortunately, the weather during the 1890s did not severely impact the city to cause disasters while construction was in progress. Heavy rains in the spring of 1892 brought this potential and helped to urge the awarding of the first contracts later that year. Another storm in March 1894 was reported to have caused damage due to a failed levee near Summit. The SDC investigated and determined this claim to be unfounded. For the remainder of the decade, the levee, the Des Plaines River Dam, and the Des Plaines River diversion channel saved the city from the threat of Des Plaines River floods.

Construction Inspection and Testing

It was not until 1894 when cement masonry was used for retaining walls and bridge abutments that the issue of cement quality came about. The finding of defective cement resulted in an order to test all deliveries starting in August. The following year, cement inspectors were hired and a testing laboratory established. In 1898, more space was required for the laboratory, and the problems with contractors attempting to use defective product grew so pronounced that the board authorized the SDC to purchase cement separately and supply it to the contractors. The use of iron, steel, and other metals in bridges also brought the need for engaging the services of independent testing agencies for mill and shop inspections.

Construction Verification

Chief Engineer Randolph recommended and was authorized in February 1895 to prepare detailed estimates of all work. This occurred before the start of the fourth season of construction, and Randolph was eager to have an independent verification of the quantities used for contractor payments. To obtain independent yet qualified engineers, the SDC turned to the Western Society of Engineers in Chicago to name six candidates to lead the work for remeasurement of the Main Channel. After selection and engagement of the special engineer, work began in June and was completed and reported to the board in October. The remeasurement confirmed the original estimates within 1 percent. Additional remeasurements were made the following year on completed work also. Randolph's concerns appeared to be satisfied as no further remeasurements were performed.

Water Quality Impact on the Illinois and Mississippi Rivers

In the late 1890s, as construction of the Main Channel neared completion, the potential for its opening raised concerns for the impact on downstream water quality. The SDC invited Dr. Arthur R. Reynolds of the Chicago Board of Health to investigate and report on this matter in November 1898. By the following March, Dr. Reynolds submitted his proposal for measurement of water quality both before and after the opening of the Main Channel. The board accepted the proposal and advanced funds to Dr. Reynolds in July to begin work. A distinguished team of scientists was assembled by Dr. Reynolds for this work.

Progress reports were submitted to the SDC in 1900. In June 1900, there was some doubt as to the ability of the team to complete the study, but completion occurred by October. The results of the study were published in 1902, and the study concluded that the sewage from Chicago had been completely assimilated before the water reached Peoria. It also recognized the vastly improved quality of Chicago River waters, the improvement in the fish resources of the Illinois River, and the improvement in the quality of the Illinois River at its mouth upstream from Saint Louis.

Raising Revenue

Having sufficient revenue to carry out the requirements of the act was of concern to the board from the beginning of the SDC in 1890. For the first two years, the only revenue necessary was for paying salaries and office expenses. Finding this revenue was accomplished by the common practice of borrowing against the anticipation of tax revenue. The annual tax levy was set at the maximum rate of 0.5 percent of the value of the taxable property to develop a fund for construction. The first ordinance for a bond issue was passed in August 1892, the same month construction started on the Main Channel. The $2,000,000 bond sale was completed in November. Nine bond sales in amounts ranging from $2,000,000 to $4,000,000 were made through 1899. Indebtedness of the SDC was limited in the act at 5 percent of the value of taxable property, with a maximum of $15,000,000. The board also levied the annual ad valorem tax at the maximum rate each year. This practice provided adequate funds for construction.

In 1895, it was realized that the statutory tax rate maximum would not produce adequate annual revenue, and it was decided to seek an increase and a special committee of the board was designated to carry this message to Springfield. The statute was amended that year, raising the limit from 0.5 percent to 1.5 percent for three years: 1895, 1896, and 1897. In December 1896, it was decided to seek authority to extend the limit for two more years. It was necessary to seek this authorization early, for the general assembly only met in the odd years. The legislation was successful in 1897, and the maximum rate was continued through 1899. The SDC was able to fund more construction with current revenues and to reduce the amount funded through borrowing.

Relations with the Federal Government

It was the dream of some that the federal government would build or at least fund the building of the Main Channel, but in 1890 there was no dominant interest in federal civil works. Harbor and navigation improvements were implemented by the federal government primarily for military, not commercial, reasons. Thus, the federal government, through the US Army Corps of Engineers, improved Chicago Harbor to serve Fort Dearborn and to provide shelter for military vessels on Lake Michigan. However, in the 1890s, the federal interest in commercial navigation came about to improve and control harbors and connecting rivers. Finally, the 1899 Rivers and Harbors Act

CHAPTER 11: ANCILLARY ISSUES

established federal dominance for control and improvement to the benefit of commercial navigation.

In August 1892, after an inspection of the channel route and review of the plans of the SDC, the corps lauded the SDC undertaking in a letter to the board. The SDC requested federal participation and was immediately rebuffed in an exchange of letters in January 1894. Later, in April, came a report of failure to obtain a congressional appropriation. Corps generals from Washington, DC, inspected the work in 1895. In February 1898, the Rivers and Harbors Committee of the House of Representatives was invited to inspect the Main Channel work, and the inspection occurred in June, but no funding was promised. The Main Channel was built without federal funding or oversight. However, late in the decade, federal control began to be felt. As explained in chapter 7, improvement work in the South Branch required federal permits issued by the Corps, as did the eventual connection to the South Branch and the opening of the Main Channel.

Relations with the I&M Canal Commissioners

On a mission to build a canal near the I&M Canal and to cross it in Joliet, there were bound to be problems between this fifty-year-old agency and the new SDC. Many problems regarding the selection of the Main Channel route were causes for friction with the I&M Canal Commissioners. The issues dealing with route selection and construction problems involving the I&M Canal are explained in chapters 4 through 8. There were more, however. In May 1893, the commissioners made a demand on the SDC for the raising of two bridges crossing the I&M Canal, which the SDC now owned by virtue of purchasing the land next to the I&M Canal. The commissioners wrote to the US Congress in November 1893, wanting very much for the I&M Canal to be designated the link in the lake to Gulf Waterway. The SDC paid for repairs to the I&M Canal banks on several occasions as a result of damages caused by SDC contractors. The damages included lost tolls due to canal traffic interruption. As completion of the Main Channel neared, the SDC agreed to pump water into the canal to maintain navigable depth.

Relations with the City of Chicago

Given the significance that reversal of the river would have on the city, it is surprising that there was not more official interaction and cooperation between the two bodies. In July 1895, the city inquired regarding the status of construction of the Main Channel, when the channel would be opened, and plans for additional channels and sewers. This inquiry led to a conference involving the mayor, the commissioner of public works, and the SDC in November 1897. The city was seeking SDC construction of relief sewers, but the SDC declined, limiting its work to the conveyance of wastewater away from the city. The SDC did agree to be involved in the Thirty-Ninth Street and Lawrence Avenue Conduits, which conveyed sewage from lakefront intercepting sewers to the river. Another issue was the drainage of the area south of Seventy-Fifth Street. This area was deemed too far from the Main Channel, and a separate sanitary district was sought by the city. No action occurred on the separate sanitary district, and the area was eventually served by city sewers. The SDC agreed to maintain a pumping station draining to the Main Channel. The area south of Eighty-Seventh Street was eventually drained to the Calumet River and Little Calumet River. The area between Seventy-Fifth and Eighty-Seventh Streets drained to the north and connected to the Main Channel.

Communications and Travel along the Main Channel

The route of channel construction was paralleled by a railroad offering intercity passenger service between Chicago, Joliet, and cities downstate. Although the railroad was used extensively by SDC survey crews, the results of their work would not be known until the crew returned to the SDC office in Chicago. Horseback or horse-drawn wagons were the modes of travel other than by railroad. Once construction started, off-site horseback and railroad proved to be a deficient means of travel. In July 1893, the chief engineer was ordered to submit plans for a road along the channel to facilitate travel along the route for contractors and SDC personnel. It was recommended that the I&M Canal tow path be improved for use as a roadway. An agreement was entered into with the I&M Canal Commissioners. The tow path was improved and made available for use as a roadway between Summit and Lockport by the start of construction in the spring of 1894. As contracts were awarded between Summit and Chicago, the road was extended.

The horses used by the Police Department for patrol and the Engineering Department for construction inspection were under the control of the Police Department and stabled at the police stations. Engineers often found themselves without a horse for their work because the police took care of their own needs first. In May 1894, this problem was resolved through the acquisition of additional horses and the reservation of their use for the Engineering Department.

The SDC Board and department heads felt the need for a more immediate form of communication than receiving messages from the construction offices by messenger. The SDC entered into a contract in June 1894 with the Chicago Telephone Company for a line along the Main Channel route connecting each Engineering Department field office and police station. Contractors who wanted to connect their field offices to the Chicago exchange could do so by agreement with the SDC and payment of a monthly fee. The line was extended to Joliet in July 1897 for the work on the Des Plaines River Improvement Project.

Removal of Obstructions in the Des Plaines and Illinois Rivers

Although this topic was part of the title of the enabling legislation, little attention was paid to it by the SDC. The obstructions referred to dams in the river that would cause backup of river water and flooding as a result of the increased flow in the river when the Main Channel was opened. While increased flooding was a distinct possibility for the intended flow of 10,000 cfs, the seriousness of the problem at lesser flows was not known. In late 1897, as hope for the completion of the Main Channel was developing, the board authorized the preparation of plans for the removal of the dams.

Which specific dams needed attention was not known, so it was not until June 1899 that the board authorized investigations to be made of the dams at Kampsville and La Grange, Illinois, both downstream of Peoria. Downstate interests had other concerns and prevailed upon the SDC to investigate dams farther upstream at Copperas Creek and Henry. The investigation was carried out, but actual work did not occur until well after the Main Channel went into operation. Numerous lawsuits were filed against the SDC by downstate landowners along the river, claiming that increased flow caused damage to their property. These lawsuits dragged on well into the twentieth century.

References

Illinois Laws. Illinois, 1895 and 1897.

Reynolds, Arthur R. *Report of Streams Examination*. Chicago: SDC, 1902.

SDC. Maps – Illinois Valley. Illinois Valley Engineer. 1917.

SDC. Proceedings of the Board of Trustees of the SDC. 1890–1900.

CHAPTER 11: ANCILLARY ISSUES

Photograph 11.1 taken in 1895 shows SDC staff and Police Officers outside of a field office. MWRD photo, Geiger set, image 97.

Photograph 11.2 taken in 1895 shows SDC engineers outside of a field office. MWRD photo, Geiger set, image 188.

Photograph 11.3 taken in the mid-1890s shows a typical Rock Section contractor compressed air and steam power plant to the left and kitchen and housing buildings to the right. MWRD photo, disc 7, image 20.

Photograph 11.4 taken in 1895 shows buildings in the Section 7 contractor's construction camp. Most workers lived on site due to the lack of roads and public transportation. Each construction contractor's camp was often composed of one dominant ethnic group. MWRD photo, Geiger set, image 141.

Photograph 11.5 taken in 1895 shows an innovative use of spoil rock for local housing or shelter in Section 7. MWRD photo, Geiger set, image 291.

Photograph 11.6 taken in 1895 shows the popularity of the project to reverse the flow of the Chicago River. Here we see a group of thrill seekers in a hopper under a cableway high above the excavation of the Main Channel in Section 6. Many attendees to the 1893 Columbian Exhibition in Chicago made an effort to see this project also. Numerous tours by engineers and public officials occurred over the years. MWRD photo, Geiger set, image 127.

Chapter 12

Placing the Main Channel in Operation

The construction of the Main Channel and the improvements to the South Branch and Des Plaines River were well underway and some parts were completed as of the summer of 1898. The board was undoubtedly eager to initiate the process of obtaining the governor's approval. This process was governed by strict requirements in the act, which appeared to be straightforward but would prove difficult to implement.

Requirements of the Act

Section 27 of the act established the conditions under which the Main Channel could be put in operation. It required that the board notify the governor before any water or sewage was admitted to the Main Channel and that the construction of the Main Channel was completed. The governor would then appoint a commission of three persons. The persons were to be residents of specific areas of the state as follows: one from Joliet or of the area between Joliet and LaSalle, one from LaSalle or the area between LaSalle and Peoria, and one from Peoria or of the area between Peoria and the mouth of the Illinois River. These three commissioners were to then meet in Chicago and to appoint a competent civil engineer and such other staff as may be necessary to expeditiously perform their duties.

The commission's duty was to make such examinations and surveys of the Chicago River and the Main Channel to enable them to determine if the channel is of the capacity required by the act. If their determination was positive, they were to so certify to the governor. If their determination was negative, then they were to file a lawsuit against the SDC specifying the deficiencies in the construction. The court was directed to enjoin the SDC from admitting water or sewage to the Main Channel. The injunction was to continue until all deficiencies were resolved to the satisfaction of the court. The commissioners and staff would be paid for their services from the state treasury, and the SDC would reimburse the state.

As outlined by the act, approval would be in the control of downstate interests with the court directing any remedies thought necessary by the commission. There was no imperative for the commission to act in a timely manner and no provision for intervention by the governor. Further, the legislature must appropriate funds for the operation of the commission and its staff. While the board never took a public position on the approval process, it must have appeared to them as fraught with political and administrative pitfalls.

Initiating the Approval Process

In December 1898, drafts of two differently worded orders to notify the governor of the practical completion of work and to request appointment of the special commission were considered by the board and referred to committee. One order was returned by the committee, approved in January 1899, and sent off to Governor Tanner. The order specifically referred to the need to obtain the approval of the governor in order to turn water into the Main Channel, completion of the Main Channel in conformance with Section 23 of the act, the completion of improvements in the Chicago River and Des Plaines River, the need to appoint three commissioners to inspect the work, and the willingness of the SDC to reimburse the state for the costs of inspection.

This communication and order apparently had the desired effect of moving the governor to appoint the three commissioners, but the general assembly did not appropriate the funds needed to carry out the work of the commission. In June 1899, a special meeting of the board was called on a Saturday to consider action needed to remedy the lack of state funding. Responding to a letter from the commissioners, a board committee recommended that the SDC deposit $25,000 to be used by the commission to meet salaries and pay for other expenses in making the inspection of the Main Channel required by the 1889 act. Another needed remedy was to pay for the services of the commission's chief engineer. The act limited this compensation to $10 per diem, and the commission could not find a qualified person to take the job at this salary. The SDC Board readily agreed, and the necessary funds were deposited.

The work of the commission proceeded, and the SDC was kept apprised of the progress by the submission of payrolls and expenses. For July through December 1899, the expenses totaled $13,500.

Report of the Special Commission

A preliminary report was submitted by the commission to the SDC in November 1899, setting forth the conclusions reached based on the investigations made thus far. The commission intended to continue their work, expecting that the SDC would, at some point, announce completion of the work and turn over the channel to the commission for completion of their investigation. However, time was running out for the SDC. Not only was the pressure mounting from the public for relief from the nuisance of the Chicago River, but there was increasing debate over the impact of the Main Channel on interests downstream and upstream. The sooner approval was received, the better for water to flow through the Main Channel. Thus, the SDC seized upon the opportunity that this preliminary report presented to move forward with the approval process.

The report set forth two propositions that the commission believed were derived from the act: that the flow in the channel is required to be 5,000 cfs or 333.33 cfs per 100,000 population when the population exceeds 1,500,000 and that the required flow be maintained at all times. It went on to cite verbatim Sections 20, 23, and 25 of the act as the statutory requirements the special commission were to verify. The commission conducted a survey and found that the population of the SDC was 1,800,000, which they concluded set the required flow at 6,000 cfs.

Section 27 of the act was cited as outlining the jurisdiction of the commission to determine if the construction of the Main Channel and improvement of the Chicago River and Des Plaines River met the requirements of the act. The commission stated that, based on its investigations, the Main Channel, when freed of obstructions remaining, will meet the depth, velocity, and discharge capacity requirements of Section 23 of the act. Further, the Chicago River, when freed of obstructions, will meet the discharge capacity required to deliver flow to the Main Channel. The commission questioned whether the velocity limit in the federal permit should be considered in the capacity issue. Their counsel rendered a legal opinion finding that conditions imposed by the federal government were not within the scope of authority of the commission and should not be considered.

A detailed list of deficiencies and unfinished work was included beginning with the finding that the SDC had not implemented any means to comply with the act's requirement to provide for the removal of garbage and dead animal parts from sewage before the sewage was discharged into the Main Channel. Unfinished work in the Chicago River included removal of cofferdams at Adams and Van Buren Streets, excavation within the Bypass Channel at Van Buren Street, erection of the Taylor Street Bridge, removal of the center pier at the Chicago Terminal Railroad crossing, construction of abutments and erection of the new Chicago Terminal Railroad Bridge, and widening of the South Branch downstream of the Eighteenth Street Bridge. The latter was included, although it was acknowledged as a federal project.

Unfinished work in the Main Channel was identified. Section O needed removal of the dam near Western Avenue, completion of masonry abutments and removal of temporary piling for the Eight-Track Bridge, completion of Main Channel and Collateral Channel excavation, and construction of the flume for admitting water into the Main Channel. Section N needed completion of excavation and removal of a temporary dam. Section K needed construction of a temporary trestle, center pier, and abutments for the Belt Line Railroad crossing and completion of channel excavation. Section H needed removal of two temporary dams. Section 8 needed completion of rock excavation in the Main Channel bottom. The commission acknowledged that work on all these items was underway.

Work to be completed on the improvement of the Des Plaines River included Dam No. 1, rock excavation, and construction of bridges in Joliet. The report also referred to the extensive amount of cleanup work remaining and the required removal of the Illinois River Dams at Henry and Copperas Creek. Last, the commission suggested that a water-flow gauge be installed at Lockport to verify that the proper amount of water was being discharged. They ended the report by reiterating their desire to see the work completed and requesting that the SDC complete the work and turn the Main Channel over to the commission before they reported to the governor. The prospects for a speedy resolution and approval were not apparent in the language of the commission's preliminary report.

Response by the SDC

The commission's report was delivered to the SDC Board at a meeting of the two bodies in November. The SDC responded at the meeting to some of the issues raised in the report. In particular, the SDC indicated that the demand to complete all the unfinished work before the commission could report to the governor would delay for several months the opening of the Main Channel and cause a great hardship on the citizens of Chicago. The SDC suggested that the commission could approve the work when the channel would pass the required discharge at the specified velocity, regardless of the existence of temporary construction work, such as cofferdams and pilings.

Two days after receipt of the commission's report, the SDC responded in writing with a firm rebuttal on some issues and commitment on unfinished work details. The SDC denied that a special population count was warranted, suggesting that it had no legal precedent and indicating that the federal census should be used absent any other direction in the act. The 1890 census counted the population in the SDC territory at approximately 1,100,000. Further, the population used should be that which was connected to the Main Channel and not the population of the SDC territory. Since many auxiliary channels and sewers have yet to be built to connect the population to the Main Channel, the discharge capacity should be no more than the statutory minimum of 5,000 cfs. The flow required by the connected population would not cause the velocity to exceed the limit in the federal permit.

In response to the commission's comment on unfinished work in the Des Plaines River through Joliet, the SDC rejected the commission's authority to include the Des Plaines River in Joliet in their investigations. This work was undertaken by the SDC to protect its own interests, was not contemplated by the act, and no requirements for it had been set forth in the act. The response also reiterated the SDC's position on moving forward with the approval process despite temporary construction work in the Main Channel.

The SDC offered commitments with a date certain for completion of each item of unfinished work mentioned in the commission's report. Completion dates ranged from the end of November to the middle of December. Additional commentary was supplied for some of the items. The erection of the Taylor Street Bridge would have no impact on the discharge in the South Branch; therefore, its completion was not relevant. The federal project to widen the South Branch downstream of Eighteenth Street was to benefit navigation and would not materially add to the cross section of the river channel. The completion of the Collateral Channel and the removal of the temporary dam near Western Avenue would be performed by dredging after the Main Channel had been filled with water. Since some of the discharge through the Main Channel would come through the Collateral Channel, the velocity of flow in the Main Channel east of the confluence would be less than required by the act; therefore the temporary construction work at the Eight-Track Bridge would not restrict the discharge.

All the work in Sections N, K, and 8, including the work on the Belt Line Railroad crossing that had just begun, would be finished by early December. The two low dams in Section H did not have to be excavated as they were designed to be removed by the flow of water and deposition of the solids was provided for in areas below each dam that were excavated below the channel invert for this purpose. The SDC had ordered all contractors to expedite the work, providing extra compensation and incentives for early completion.

Opening the Main Channel

The construction of the Main Channel and improvements to the Chicago River and Des Plaines River was now approaching completion. This project was a remarkable feat at the time. A total of 42,230,000 cubic yards of rock and soil were excavated, and 460,000 cubic yards of masonry for channel walls and bridge abutments were built. The Main Channel downstream of Summit had a capacity, as required by statute, of 10,000 cfs; and the channel upstream of Summit, excavated in earth, had a capacity of 5,000 cfs. The channel upstream of Summit would later be enlarged by dredging to achieve the larger capacity in compliance with the act. The total cost of the Main Channel and Des Plaines River work from Robey Street to Joliet was $33,530,000.

There were no further official meetings or exchanges of reports or correspondence between the SDC and the commission. Obviously, there was communication of some kind to bring about the governor's eventual approval. The 28-mile long Main Channel was without water, except for the seepage that was pumped by the contractors or allowed to flow toward Lockport, where it was pumped by the SDC to the Des Plaines River. Before the Main Channel could be placed into operation, it must be filled slowly so as not to cause damage by rapidly rising water levels or swift currents. Water was let in at the Chicago end beginning on January 2, 1900, at 10:35 a.m. The earth dike across the south end of the Collateral Channel in front of the entrance to the wooden flume was breached. The flume protected the earth slopes of the Main Channel from erosion, which would otherwise have occurred without the flume.

The letting in of water to the Main Channel would normally have been an occasion for celebration, but due to the hurried effort, there was no such plan. The board members, Chief Engineer Randolph, several staff members, and newspaper reporters were present along with the contractor as early as daybreak. The board members had wanted to move the few shovels of dirt themselves to initiate water flowing. Given the cold weather, ice, and tough clay, muscle power and shovels, even a stick or two of dynamite, were insufficient to make progress. Eventually the people present were moved aside, and the contractor's dredge was moved in to do the job of breaching the dike. Even the dredge had difficulty but soon prevailed, and the water flowed.

Four days later, at 4:54 p.m., the water level rose to the sill of the gates at the Lockport Controlling Works. All gates were closed, and the Bear Trap Dam was raised to its highest level to contain the water in the Main Channel. The filling continued to January 14 when the water level in the Main Channel reached the water level in the West Fork. The next day, the earth dam across the Main Channel west of Western Avenue was cleared away by dredges. The waters on each side came together at 11:08 a.m. After thirteen days of filling, the water level came to rest to wait for the last remaining deliberations of the commission in reporting to the governor.

In the early hours of January 17, 1900, the day of a regularly scheduled board meeting at 2:00 p.m., the SDC trustees travelled to Lockport to be at the Lockport Controlling Works when word was received from the governor. The members of the commission were also on hand after their earlier meeting and report to the governor. The governor's approval was received by telegram, and at 11:05 a.m. on January 17, 1900, the Bear Trap Dam was lowered slightly below the water level to allow a thin sheet of water to flow over the top thereof. In a brief ceremony, SDC Board President Boldenweck introduced Colonel Isaac Taylor, president of the special commission. Colonel Taylor summarized the findings of the commission, leading to their conclusion of satisfactory completion.

Immediately thereafter, the president gave a signal; the valves controlling the dam were opened, and the massive 160-foot-long dam disappeared beneath the water. At 11:16 a.m. a torrent of water rushed out of the Main Channel over the dam toward the Des Plaines River. On this chilly day in the first January of the new century, ten and a half years after passage of the act, ten years after the first meeting of the SDC Board of Trustees, and after seven and a half years of construction, the Main Channel was now in operation to save Chicago from its own waste.

The Bear Trap Dam served for many years as the regulating control for the Main Channel until replaced by the operation of the Lockport Powerhouse located 2 miles downstream in 1908. In the words of Chief Engineer Isham Randolph, "The Bear Trap Dam was successfully operated on the 17th, and was in service daily the balance of the month, performing the work for which it was designed, fully and perfectly in every respect."

Closing the Books on the Special Commission

Several months after the opening of the Main Channel, the SDC made a demand upon the commission for the return of all funds remaining on account as of January 17, 1900. The demand was based on a belief that the work of the commission was completed and that its existence terminated as of that date. If the commission refused the demand, the SDC Attorney was ordered to initiate legal proceedings to recover the funds. However, in July 1900, a majority of the board had a change of heart and voted to compensate the commission for their work up to August 1, 1900.

Colonel Taylor explained to the board that the commission was required to document their investigations and prepare a formal report to the governor to justify and support the governor's approval, which was given expeditiously upon the oral report of the commission in January. The majority report of the board went into further detail, revealing some of the intrigue behind the governor's hurried approval. At the time, a number of lawsuits had been entered or threatened to be entered against the SDC by the City of Saint Louis and others in both federal and local courts to enjoin the SDC and prevent the opening of the Main Channel.

Undoubtedly, the intrigue was based on the threat of a lawsuit in federal court by Saint Louis that could result in enjoining the governor from granting his approval or the SDC from placing the Main Channel in operation. As others have explained, Saint Louis was seeking approval to be the site of the 1904 World's Fair and needed the approval of the US Congress. The Illinois congressional delegation and the delegations of other states could block this approval if Saint Louis persisted in trying to block the opening of the critical link in an interstate navigable waterway linking the Great Lakes and the Gulf of Mexico. Saint Louis eased off the lawsuit and got the Fair.

The commission took a generous view of the situation. Rather than hold up the approval process on the basis of the technicalities of incomplete construction and subject the whole undertaking to indeterminate delay in the courts, the commission urged the governor to approve of the opening of the Main Channel to relieve the citizens of Chicago of the sanitary nuisance, to insure the safety of the drinking water supply, and to create the waterway link connecting the Great Lakes with the

Gulf of Mexico. There was a tacit understanding at the time that the commission would be allowed to take the time needed to make a thorough report. For these reasons, the funding of the commission to August 1 was justified.

References

Chicago Tribune, January 3, 1900.

Engineering Works. SDC, 1928.

Report of the Special Commission. Special Commissioners Chicago Drainage Channel. Illinois, 1900.

SDC. Proceedings of the Board of Trustees of the SDC.1898, 1899, and 1900.

Williams, Michael, and Richard Cahan. *The Lost Panoramas: When Chicago Changed Its River and the Land Beyond*. Chicago: CityFiles Press, 2011.

CHAPTER 12: PLACING THE MAIN CHANNEL IN OPERATION

Photograph 12.1 taken on December 1, 1899 shows the location where water from the West Fork of the South Branch was introduced in Section O to fill the excavated Main Channel. A wooden flume, seen in the center, was constructed in the northwest bank of the Main Channel adjacent to the south end of the Collateral Channel. The Chicago, Madison and Northern Railroad Bridge is to the left and the center pier for the Kedzie Avenue Bridge is seen below the railroad bridge. Controlled filling was critical to maintain the stability of the side slopes of the Main Channel and structures. In the month of December work was completed to make the channel ready for water. This view is looking westerly. MWRD photo, disc 13, image 67.

Photograph 12.2 taken on December 30, 1899 shows the channel nearly ready to begin the filling process. The materials in the right foreground would be removed and the channel side slope graded by the Section N excavation contractor in the final few days. The filling flume is located behind the Chicago, Madison and Northern Railroad Bridge. This view, taken from the Kedzie Avenue Bridge, is looking northeasterly. MWRD photo, disc 14, image 81.

CHAPTER 12: PLACING THE MAIN CHANNEL IN OPERATION

Photograph 12.3 taken on January 2, 1900 shows the initial inflow of water from the Collateral Channel through the filling flume in Section O. The Board of Trustees and staff of the SDC turned out to witness this momentous occasion. In the right background is a dredge used to open the connection to allow water to flow. MWRD photo, disc 127, image 20.

Photograph 12.4 taken on January 2, 1900 shows the lineup of nine members of the Board of Trustees and Chief Engineer Isham Randolph, second from left, ready to take the first shovel full of earth to allow water to flow into the Main Channel. MWRD photo, disc 14, image 75.

CHAPTER 12: PLACING THE MAIN CHANNEL IN OPERATION

Photograph 12.5 taken on January 2, 1900 shows the dipper dredge of the excavation contractor in Section O breaching the earthen berm at the south end of the Collateral Channel to allow water to flow into the Main Channel through the filling flume. MWRD photo, disc 14, image 77.

Photograph 12.6 taken on January 9, 1900 shows the slowly rising water level at the Bear Trap Dam of the Lockport Controlling Works in Section 15. A crowd has gathered to witness the rise. As of this date, the Main Channel is slightly more than half full. The full height of the Bear Trap Dam is shown by the shading on the far wall. MWRD photo, disc 14, image 73.

Photograph 12.7 taken on January 16, 1900 shows the Main Channel full in Section 12. This view is taken from the Romeoville Road Bridge looking north. The water level in the Main Channel reached equilibrium with the water level in the West Fork and Collateral Channel on January 14, 1900. The wait began for Illinois Governor Tanner to give his approval for the water to flow out of the Main Channel at Lockport. MWRD photo, disc 14, image 61.

Photograph 12.8 taken on January 16, 1900 shows the water in the Main Channel at Lockport at its maximum level in equilibrium with the water level in the West Fork of the South Branch in Chicago. The Bear Trap Dam is raised to its highest level to contain the water. Governor Tanner gave his approval to allow water to flow on January 17, 1900. MWRD photo, disc 14, image 59.

Photograph 12.9 taken on January 20, 1900 shows the flow of water over the Bear Trap Dam from the Main Channel into the tailrace channel and on to the Des Plaines River three days after the dam was lowered and flow began. The flow in the Chicago River was now irrevocably reversed. MWRD photo, disc 14, image 49.

Epilogue

The Main Channel and the remainder of the Chicago Area Waterway System (CAWS), after more than a hundred years of successful service, is facing new challenges. The Main Channel has evolved into the CAWS, and many changes have taken place over the years to control and improve flow and capture renewable energy: extension of the Main Channel; relocation, widening and deepening of the North and South Branches; widening of the Earth Section; addition of the man-made North Shore Channel and Calumet-Sag Channel; deepening and widening of the Chicago and Little Calumet Rivers; and installation of gates, generators, locks, pumps, and turbines. Other enhancements include widening of the Calumet-Sag Channel for commercial navigation, construction of water reclamation plants for sewage treatment, construction of deep tunnels for additional combined sewer overflow storage, and supplemental aeration to improve water quality.

The CAWS has matured into a 77-mile network of artificially controlled man-made canals. The water in the CAWS is manipulated by the gates, locks, pumps, and turbines. These devices are controlled and operated by the MWRD. Safety on the water is under the control of the US Coast Guard. Construction in the water must be permitted by federal, state, and sometimes local authorities. Most of the network is part of the Illinois Waterway, a federal navigation project under the control of the US Army Corps of Engineers. The CAWS is also the Lockport Navigation Pool, the most upstream pool in a series of eight pools on the Illinois Waterway between Grafton and Chicago.

The Lockport Pool or CAWS is near static, with the water surface between the Lockport Lock and the Chicago River Lock, some 35 miles apart, being only 2 or 3 inches different in elevation on most days. Water flow between these two points is slow, measured in days and weeks. When it rains, the MWRD must convert the CAWS immediately from a static into a dynamic system to effect drainage of the Chicago area. This is accomplished by manipulating the gates, locks, pumps, and turbines to increase the rate of flow at Lockport, decrease inflow from the lake, and lower the water level to create storage for incoming storm water.

When all floodwater cannot be effectively discharged in a timely manner through Lockport, it becomes necessary to discharge excess floodwater into Lake Michigan at one or more of the three lakefront control points. Added storage in underground tunnels and below ground reservoirs, when available, has and will reduce the frequency of discharges to the lake. However, predictions based on climate change indicate that rainfall frequency and intensity, hence flooding, may become more severe.

In 2011, the designated uses and water quality standards of the CAWS under the Clean Water Act (CWA) are at issue before the Illinois Pollution Control Board. Aquatic life in the CAWS, also subject to protection under the CWA, is part of this regulatory issue where there is uncertainty about the interdependence of aquatic habitat and water quality in this man-made waterway. The quality of permitted discharges to the CAWS under the CWA is in contention between federal and state permit-issuing authorities and numerous dischargers. The movement of aquatic invasive species between the Great Lakes and the Mississippi River watersheds has brought focus on the potential for separating the long-standing hydraulic connection between these two basins. The conversion of industrial land along the CAWS to commercial and residential uses raises concern for aesthetics and riparian amenities. Increased recreational use of the CAWS brings recognition to the need for public health protection.

There are other lingering issues that have yet to grab the attention of the public: past industrial and storm water inputs have left contaminated sediments; careless disposal of unwanted materials have left unsafe bottom surfaces containing rocks, glass, metal, and soft ooze; dock walls, fences, sheet piling, and steep banks restrict safe ingress and egress to the water for recreation; societal dependence on new products for better living leaves uncounted and unregulated chemicals in treated wastewater and storm water; and our collective and careless use of paper and plastic products allows litter to blow into the water. Even our desire for biodiversity and wildlife in and out of the water raises questions about the impact of nonhuman microbial pathogens and pollutants.

Maintaining the quality of Lake Michigan as a reliable source of water for the Chicago area was what motivated the building of the Main Channel in the 1890s. This construction altered the surface drainage system in a way that seems irreversible. In addition to all the changes to the CAWS that have occurred since then, further changes will undoubtedly occur. Whatever becomes of the CAWS, the quality of Lake Michigan and efficient drainage will be essential to the continued vitality of the Chicago area. How we manage our water going forward is critical to a sustainable future.

Acknowledgments

While I alone take responsibility for accuracy and content, I could not have completed this history without the assistance and support of many. My wife, Marsha Richman, has been patient beyond belief with my pursuits of details and my many hours on my desktop or laptop computers. Libby Hill provided much inspiration and example of bringing a history to completion with her book, *The Chicago River: A Natural and Unnatural History*, and it was my pleasure to assist her with research and editing and have her counsel on my manuscript.

In 1967 I succeeded Don R Brown as Engineer of Waterways Control. Don began his career at the SDC in 1924 when lake diversion was at its peak and the principal business at SDC was the generation and transmission of electricity. SDC electricity from the Lockport Powerhouse operated SDC pumping stations and illuminated boulevard and streets in Chicago. I learned about the early days of diversion and sewage treatment innovation at the SDC from Don and other SDC seniors like Norval Anderson, Phil Furlong, Ted Ladd, Ted Mickle, Leo Peller, Milt Pollock, John Towne and Bill Wheeler.

My recent interest in MWRD history has benefitted from numerous members of the staff at the MWRD who have supplied helpful insight and access to records that although public, might not have been easy to locate. Histories written by others; photographs; engineering records and reports; drawings and maps; operational logs, records, and reports; and many more resources have defined, explained and illustrated my understanding. I'm deeply grateful to the assistance given by many too numerous to mention.

The governance of the MWRD is entrusted to the board of commissioners and without the resources provided by the board, the staff would not be able to serve the public as well as they do. I'm grateful to the board members, past and present, for making the critical decisions on policy, programs and resources.

Bridge List Appendix

Numerous bridges were necessary to effect the reversal of the flow of the Chicago River, and these were built by the SDC, except where noted. The bridge substructure and superstructure contracts were normally awarded to different contractors at the same time. The year of completion is indicated or, if not completed, the extent of completion in percent at the end of 1899. The bridges are listed from Chicago to Joliet.

CHICAGO RIVER IMPROVEMENT

VAN BUREN STREET APPROACH SPAN OVER THE BYPASS CHANNEL

Fixed plate-girder type, double roadway each 18 feet wide, two sidewalks each 8 feet wide, 73.3 feet in length. Completion: 43 percent.

TAYLOR STREET OVER THE SOUTH BRANCH

Single-leaf bascule, rolling lift bridge, Scherzer type, single roadway 18 feet wide, two sidewalks 5 feet wide each, 148.6 feet between bearings, channel clearance of 120 feet, weight of 500 tons. Completion: 39 percent.

CHICAGO TERMINAL RAILROAD OVER THE SOUTH BRANCH

Single-leaf bascule, rolling lift bridge, Scherzer type, double track, 275 feet between bearings, channel clearance of 120 feet, weight of

2,445 tons. Completion: 23 percent.

EARTH SECTION

SOUTHWEST BOULEVARD AND WESTERN AVENUE OVER THE MAIN CHANNEL IN SECTION O

Center-pier swing bridge, two roadways 24 feet wide each, two sidewalks 6 feet wide each, 321 feet in length, 729 tons in weight. One roadway was for boulevard traffic, and the other was for avenue traffic. Completed in 1899.

EIGHT-TRACK RAILROAD CROSSING OVER THE MAIN CHANNEL IN SECTION O

Three railroads shared four identical, parallel bridge spans. Each bridge is a single-leaf bascule, Scherzer type, double track, 150 feet between bearings, with hinges on alternate sides of the channel. Each bascule has two approach fixed plate-girder spans on each end, 113 feet between bearings. Total length of each bascule plus approach spans is 454.3 feet. Total weight of all spans is 3,038 tons. Completion: 47 percent.

MADISON AND NORTHERN RAILROAD OVER MAIN CHANNEL IN SECTION N

Center-pier swing bridge, double track, 479.4 feet in length, 1,256 tons in weight. Completed in 1899.

MADISON AND NORTHERN RAILROAD OVER KEDZIE AVENUE IN SECTION N

Fixed plate girder, double track, 80 feet in length, 100 tons in weight. Completed in 1899.

KEDZIE AVENUE OVER THE MAIN CHANNEL IN SECTION N

Center-pier swing bridge, single roadway 21 feet wide, two sidewalks 5 feet wide each, 324.5 feet in length, 320 tons in weight. Completed in 1899.

SANTA FE RAILROAD OVER MAIN CHANNEL IN SECTION N

Center-pier swing bridge, double track, 327.7 feet in length, 760 tons in weight. Completed in 1899.

BELT LINE RAILROAD OVER THE MAIN CHANNEL IN SECTION K

Center-pier swing bridge, quadruple track, 340.2 feet in length, 1,346 tons in weight. Completion: 34 percent.

SANTA FE RAILROAD OVER THE MAIN CHANNEL IN SECTION G

Center-pier swing bridge, double track, 372.5 feet in length, 862 tons in weight. Completed in 1898.

EARTH AND ROCK SECTION

SANTA FE RAILROAD OVER THE DES PLAINES RIVER IN SECTION F

Fixed plate-girder bridge, double track, six spans, 312.6 feet in total length, 250 tons in weight. Built by the Santa Fe with 50 percent of the cost paid by the SDC per agreement. Completed in 1898.

SUMMIT-LYONS ROAD OVER THE MAIN CHANNEL IN SECTION F

Center-pier swing bridge, single roadway 18 feet wide, 323.8 feet in length, 185 tons in weight. Completed in 1899.

SUMMIT-LYONS ROAD OVER THE DES PLAINES RIVER IN SECTION E

Fixed truss bridge, single roadway 18 feet wide, one span, 200 feet in length, 57 tons in weight. Completed in 1898.

CHICAGO TERMINAL RAILROAD OVER THE MAIN CHANNEL IN SECTION E

Center-pier swing bridge, double track, 316.6 feet in length, 526 tons in weight. Completed in 1898.

CHICAGO TERMINAL RAILROAD OVER THE DES PLAINES RIVER IN SECTION E

Fixed truss bridge, double track, one span, 105 feet length between bearings. Completed in 1899.

ROCK SECTION

WILLOW SPRINGS ROAD OVER THE MAIN CHANNEL IN SECTION 1

Bobtail swing bridge, single roadway 20 feet wide, 306.1 feet in length, 170 tons in weight with a counterweight of 105 tons. Completed in 1899.

SANTA FE RAILROAD OVER THE DES PLAINES RIVER IN SECTION 8

Fixed plate-girder bridge, double track, 12 spans, 720 feet in total length, 559 tons in weight. The Santa Fe Railroad built the substructure, and the SDC built the superstructure. The SDC paid 50 percent of the total cost per agreement. Completed in 1898.

SANTA FE RAILROAD OVER THE MAIN CHANNEL IN SECTION 8

Bobtail swing bridge; double track; 398.5 feet in length; 1,158 tons in weight with a counterweight of 429 tons. Completed in 1899.

STEPHENS STREET OVER THE MAIN CHANNEL IN SECTION 8

Bobtail swing bridge, single roadway 20 feet wide, 306.1 feet in length, 170 tons in weight with a counterweight of 104 tons. Completed in 1899.

SANTA FE RAILROAD OVER STEPHENS STREET IN SECTION 8

Fixed plate girder, double track, 60 feet in length, 80 tons in weight. Completion: 84 percent.

LEMONT ROAD OVER THE DES PLAINES RIVER IN SECTION 8

Fixed timber beams on timber trestles, single roadway 18 feet wide, 10 spans, 500 feet in total length. Built by the excavation contractor as extra work. Completed in 1894.

WESTERN STONE CO. RAILROAD OVER DES PLAINES RIVER IN SECTION 10

Fixed timber beams on timber trestles, single track, 32 spans, 512 feet in total length. Built by the excavation contractor as extra work. Completed in 1894.

ROMEOVILLE ROAD OVER THE MAIN CHANNEL IN SECTION 12

Bobtail swing bridge, single roadway 20 feet wide, 306.1 feet in length, 170 tons in weight with a counterweight of 104 tons. Completed in 1898.

DES PLAINES RIVER IMPROVEMENT

LOCKPORT ROAD OVER THE DES PLAINES RIVER IN SECTION 16

Fixed through truss, single roadway 18 feet wide, three spans, 600 feet in total length, 166 tons in weight. Completed in 1899.

WIRE MILLS ROAD OVER THE DES PLAINES RIVER IN SECTION 16

Fixed through truss, single roadway 18 feet wide, three spans, 500 feet in total length, 130 tons in weight. Completed in 1899.

JOLIET AND EASTERN RAILROAD OVER THE DES PLAINES RIVER BETWEEN SECTIONS 16 AND 17

Fixed through truss and plate-girder bridge, single track, one through truss span and four girder spans, 660.3 feet in total length, 557 tons in weight. Completed in 1897.

TOWPATH OVER THE DES PLAINES RIVER IN SECTION 17

Fixed truss bridge, single roadway 12 feet wide and single towpath 6 feet wide, three spans, 624.3 feet in total length, 222 tons in weight. The I&M Canal entered the Des Plaines River from the east and exited the river to the west. This bridge provided for the towed canal traffic to cross the river current. Completion: 28 percent.

CASS STREET OVER THE DES PLAINES RIVER IN SECTION 18

Fixed through truss and plate girder, single roadway 30 feet wide and two sidewalks 7 feet wide each, one truss span and one girder span, 303.4 feet in total length, 418 tons in weight. Completion: 89 percent.

JEFFERSON STREET OVER THE DES PLAINES RIVER IN SECTION 18

Fixed through truss, single roadway 37.5 feet wide and two sidewalks each 11 feet wide, two spans, 227.8 feet in total length, 326 tons in weight. Completion: 40 percent.

ROCK ISLAND RAILROAD OVER THE DES PLAINES RIVER IN SECTION 18

Fixed skewed plate-girder bridge, quadruple track, two parallel double track spans, 159 feet in total length, 159 tons in weight. Completion: 30 percent.

Reference

SDC. Proceedings of the Board of Trustees of the SDC for 1894 and 1897 through 1899.

Glossary

act: An act to create sanitary districts and to remove obstructions in the Des Plaines and Illinois River.

Belt Line: Chicago & Western Indiana Railroad Company.

board: Board of Trustees of the SDC.

Alton: Chicago & Alton Railway Company.

CCD: Chicago City Datum.

cfm: Cubic feet per minute.

cfs: Cubic feet per second.

Chicago: City of Chicago.

Chicago Terminal: Chicago Terminal Transfer Railroad Company.

Fort Wayne: Pittsburgh, Fort Wayne & Chicago Railway Company.

I&M Canal: Illinois and Michigan Canal.

JCEF: SDC Board Joint Committee on Engineering and Finance.

Joliet and Eastern: Elgin, Joliet and Eastern Railroad Company.

Junction: Chicago Junction Railway Company.

Madison and Northern: Chicago, Madison and Northern Railroad Company.

Main Channel: The channel constructed between Robey Street and Lockport to reverse the flow of the Chicago River.[1]

Northern Pacific: Chicago and Northern Pacific Railroad Company.

Pittsburgh and St. Louis: Pittsburgh, Cincinnati, Chicago & St. Louis Railway Company.

Rock Island: Chicago, Rock Island and Pacific Railway Company.

ROW: Right of way.

Santa Fe: Atchison, Topeka and Santa Fe Railway Company.

Scherzer: Scherzer Rolling Lift Bridge Company.

SDC: Sanitary District of Chicago.

Stock Yards: Union Stock Yards and Transfer Company.

West Side: Metropolitan West Side Elevated Railroad Company.

---

1 Main Channel was the commonly used name for the channel constructed from 1892 to 1900 to reverse the flow of the Chicago River. For the first few years of this period, it was also referred to as the main drainage channel. The use of Chicago Sanitary and Ship Canal appeared later in the decade on public documents and was the name used outside the SDC to promote the Great Lakes to Gulf Waterway. Oddly, it does not appear in the Proceedings of the Board during this first decade. It eventually became the official name used on maps and public documents and included the 4-mile-long Main Channel Extension built from 1903 to 1907. Main Channel continues to be used as a common name.

Index

Page numbers in italics refer to photographs and captions. The letter "t" following a page number denotes a table; the letter "f" following a page number denotes a figure.

A

Act of 1889, 15–21, 329-334
Artingstall, Samuel G., 5, 28, 29, 31, 34, 222

B

Bear Trap Dam, 54-56, 65-67, 71m, *108-112, 114, 115,* 335, 336, *344, 346, 347*
Board of Trustees of the Sanitary District of Chicago
    chief engineer and the, 26-29, 34
    elections of, 20, 30-31
    organization of, 21-22, 25-26, 37
    review of past planning of, 32-33
    strategy for the future, 33-34
Bridgeport Pumping Station, 2-3, 223-225
bridges
    deliberations on, 245-248
    in Earth and Rock Section, 125, 131m, *270, 272, 277, 279-281, 291-292, 298-300*
    in Earth Section, 159-160, 165m, *257-265,* 271, *273-277, 284-290*
    in Joliet Project, 197, 200m, 253, *204, 206, 208-209, 215, 218, 303-308*
    in Rock Section, 57, 68m-70m, *94-95, 66-69, 78, 283, 293-297, 301, 309*
    for Santa Fe System, 70m, 132m, 165m, 252, *93-95, 266-267, 278, 288, 290, 294, 298, 301, 309*
    over South Branch, 230, 233m, 253
    types of, 248-250, *284-309,* 355-360
Bypass Channel, 225-232, 233m, *235-244*

C

Carter, Orrin, 118, 222
Chicago Board of Health, 312
Chicago Portage, 2
Chicago River
    first reversal of, 2-3
    flooding of, 4
    relief routing for, 149-152
Chicago River Improvement
    bridges of. *See* bridges: of the South Branch
    construction seasons of, 231-232
    contract requirements of, 228-230
    overview of, 35m, 221-228, 255m
Citizens' Association, 4
Collateral Channel, 154-155, 157, 161-164, 165m, 332, 334-335, *339, 341-343*

Commission on Drainage and Water Supply, 4-8
Committee of Five, 8-9
Committee on Engineering, 26, 29-33, 44
contracts, awarding of, 42-43, 53, 118-119, 152-153, 189-190, 226
Cooley, Lyman E., 4-5, 25-27, 29-31, 33-34, 125, 149-150, 186-187, 246

D

Des Plaines River
    diversion channel of, 44-45, 50-51, 70m, *72*, 119, 121-123, 126, 131m, 132m, 223-224
    flooding from, 2-4, 316
    Joliet Project and. *See* Joliet Project
    removing obstructions in, 321-322
    sewage discharge into, 5-7
    tailrace channel of, 52, *106*, 186-190, 194-198, 200m, 201m, *203-204, 217, 219, 347*
Des Plaines River Spillway, 121, 132m, *146-147*
Drainage Boundary Commission, 19-20

E

Earth and Rock Section
    bridges in. *See* bridges: in Earth and Rock Section
    construction methods, 126-127, *133-145*
    construction seasons, 127-130
    contract requirements, 119-120
    contractor continuity, 126
    overview, 117
    route selection, 118-119, 131m
Earth Section
    bridges in. *See* bridges: in Earth Section
    construction innovations, 157-159
    construction methods, 160-161, *167-182*
    construction seasons, 161-164
    contract requirements, 155-157
    contractor continuity, 160-161
    overview, 149-155
    route selection, 149-152, 165m
Eckhart, B.A., 8, 30

Eight-Track Bridge, 159, 166m, 250-252, *257, 263-265, 285,* 332, 334, 356
employment issues, 313-314t, *323-324*

F

flooding, 2-3, 121, *147*, 316

G

glacial drift, 45, 52, 60, 124-125
Goose Lake, 50

H

Harrison, Carter H., 5-6
health issues, 48, 222-225, 311-312
Hughes, William, 247, 249

I

Illinois & Michigan Canal (I&M Canal)
    Joliet Project and. *See* Joliet Project
    in photographs, *11-13, 87, 207-216, 306-308*
    limited capacity, 2-3
    on maps, 10m, 23m, 36m, 68-71m, 131-132m, 165-166m, 200-202m, 233m
    planning for, 28-31, 188-191
    right-of-way of, 16
    using route of, 149-152
    as temporary relief, 222-224
I&M Canal Commissioners, 16, 151, 185, 188-191, 319, 321-322
Illinois River, 15-17, 27, 317, 321, 333
Illinois State Board of Health, 7, 224, 311

J

JCEF (Joint Committee on Engineering and Finance), 44
Johnston, T.T., 5, 49, 55, 64-65
Joliet Project
    bridges in. *See* bridges: of the Joliet Project
    construction methods, 189-193, *203-219*, 253
    construction seasons of, 197-199
    contract requirements, 193-196

overview of, 25, 42, 185-189
route selection for, 185-189, 200m-202m

L

labor relations, 312-313
Lake Michigan, XVII, 1, 3-7, 9, 16-17, 54, 56, *146*, 160, 192, 221-223, 229, 318, 350-351
legal issues, 20-21, 190-191, 337
Lockport Controlling Works, 54-56, 64-65, 68m, 71m, *105-116*, 186-187, 190, 192, 194, 198, 200m-201m, *344, 346-347*

M

Main Channel
 alternative schemes for, 40-41
 approval for operation of, 329-336
 communications and travel along, 320-321
 grade change of, 49-50
 new plan for, 29-30
Mississippi River, 2, 55, 317, 350
Mud Lake, 2-3, 156

N

Newton, John, 27-28

O

Ogden Dam, 3, 132m
Ogden-Wentworth Ditch, 3-4, 10m, 23m, 28-29, 36m, 121, 132m, *147*, 156, 165m

P

Prendergast, Richard, 19, 27, 30

R

Randolph, Isham, 43, 49-52, 55, 121, 124-125, 187-189, 191, 224-226, 231, 316-317, 335-336, *342*
referendum, 19-20
revenue, raising of, 317

Rivers and Harbors Act of 1890, 228, 318-319
Rock Island, Railroad, 189, 197, 200m, 202m, 253, *308*
Rock Section
 bridges in. *See* bridges: in Rock Section
 construction methods of, 60-62, *72-99, 103-114*
 construction seasons of, 62-67
 contract requirements of, 43-48
 contractor continuity in, 59
 controlling works. *See* Lockport Controlling Works
 earthquake damage to, 58-59, *90*
 groundbreaking of. *See* Shovel Day
 overview of, 39-43
 route selection, 40, 68m

S

SDC (Sanitary District of Chicago)
 board of trustees of. *See* Board of Trustees of the Sanitary District of Chicago
 city of Chicago and, 320
 early years of 25-34, *37*
 federal government and, 318-319
 formation of, 19-22
 I&M Canal Commissioners and, 151-152, 188-191, 319, 321
 police department of, 314-315, 321
secretary of war, 222, 225, 227-228, 253
sewage, disposal or treatment, 1-2, 4-5, 7-9, 16-17, 20-21, 224, 317, 320, 329-330, 332, 349
Shovel Day, 48-49
solutions, proposed, 3-8
South Fork, 10m, 23m, 26, 28-30, 33, 36m, 150, 233
Special Commission, 329-338
Stock Yards, 150, 159, 250
Summit-Lyons Conduit and Levee, 121-122

T

tablet, setting of, 53
Taylor, Isaac, 336-337

U

Upper Basin, 16, 189-190, 193-194, 196, 198, 202m, *205*
U.S. Army Corps of Engineers, 161, 222, 225, 318-319, 349

W

water power, 15-16, 33, 40-42, 54, 185-189, 191-192
water quality, *147*, 317, 349-350
water supply, XVIII-IXV, 1, 4-7, 17, 20, 26, 28, 34, 150, 316, 337
Wenter, Frank, 28, 30-31, 48, 53-54, 150, 224
West Arm, 28-29
West Fork, 2-4, 10m, 23m, 28, 36m, *147*, 149, 152-157, 163-164, 165m, 166m, 229, 233m, 336, 339, *345-346*
Weston, U.W., 63-64
Williams, Benezette, 5, 34, 41-43, 49-50, 52, 54, 118, 150, 186-187, 223-224
Worthen, William, 27-29